THE HISTORY OF BIOLOGY

BIBLIOGRAPHIES OF THE HISTORY
OF SCIENCE AND TECHNOLOGY
(VOL. 15)

GARLAND REFERENCE LIBRARY
OF THE HUMANITIES
(VOL. 419)

Bibliographies of the History of Science and Technology

Editors

Robert Multhauf, Smithsonian Institution, Washington, D. C.
Ellen Wells, Smithsonian, Washington, D. C.

THE HISTORY OF BIOLOGY

A Selected, Annotated Bibliography

Judith A. Overmier

GARLAND PUBLISHING, INC • NEW YORK & LONDON
1989

Library of Congress Cataloging-in-Publication Data

Overmier, Judith A., 1939–
    The history of biology.
    (Bibliographies of the history of science and
technology ; vol. 15) (Garland reference library of
the humanities ; vol. 419)
    Includes index.
    1. Biology—History—Bibliography. I. Title.
II. Series: Bibliographies of the history of science
and technology ; v. 15. III. Series; Garland reference
library of the humanities; vol. 419.
Z5320.09 1989 [QH305] 016.574'09 88-30987
ISBN 0–8240–9118–3 (alk. paper)

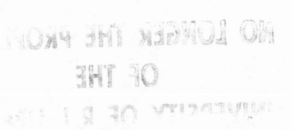

Printed on acid-free, 250-year-life paper
Manufactured in the United States of America

CONTENTS

The History of Biology

GENERAL INTRODUCTION

This bibliography is one of a series designed to guide the reader into the history of science and technology. Anyone interested in any of the components of this vast subject area is part of our intended audience, not only the student, but also the scientist interested in the history of his own field (or faced with the necessity of writing an "historical introduction") and the historian, amateur or professional. The latter will not find the bibliographies "exhaustive," although in some fields he may find them the only existing bibliographies. He will in any case not find one of those endless lists in which the important is lumped with the trivial, but rather a "critical" bibliography, largely annotated, and indexed to lead the reader quickly to the most important (or only existing) literature.

Inasmuch as everyone treasures bibliographies it is surprising how few there are in this field. Justly treasured are George Sarton's *Guide to the History of Science* (Waltham, Mass., 1952; 316 pp.), Eugene S. Ferguson's *Bibliography of the History of Technology* (Cambridge, Mass., 1968; 347 pp.), François Russo's *Histoire des Sciences et des Techniques, Bibliographie* (Paris, 2nd ed., 1969; 214 pp.), and Magda Witrow's *ISIS Cumulative Bibliography. A bibliography of the history of science* (London, 1971–; 2131 pp. as of 1976). But all are limited, even the latter, by the virtual impossibility of doing justice to any particular field in a bibliography of limited size and almost unlimited subject matter.

For various reasons, mostly bad, the average scholar prefers adding to the literature, rather than sorting it out. The editors are indebted to the scholars represented in this series for their willingness to expend the time and effort required to pursue the latter objective. Our aim has been to establish a general framework which will give some uniformity to the series, but otherwise to leave the format and contents to the author/compiler. We have urged that introductions be used for essays on "the state of the field," and that selectivity be exercised to limit the length of each volume to the economically practical.

Since the historical literature ranges from very large (e.g., medicine) to very small (chemical technology), some bibliographies will be limited to the most important writings while others will include modest "contributions" and even primary sources. The problem is to give useful guidance into a particular field—or subfield—and its solution is largely left to the author/compiler.

In general, topical volumes (e.g., chemistry) will deal with the subject since about 1700, leaving earlier literature to area of chronological volumes (e.g., medieval science); but here, too, the volumes will vary according to the judgment of the author. The topics are international, with a few exceptions, but the literature covered depends, of course, on the linguistic equipment of the author and his access to "exotic" literatures.

Robert Multhauf
Ellen Wells
Smithsonian Institution
Washington, D.C.

INTRODUCTION

The citations in this bibliography cover secondary studies in the history of biology. The bibliography is intended as an introductory level reference tool aimed at students, at faculty crossing into a new research area, and at the librarians to whom novices in the field turn to for assistance. Because of this, several general limitations are necessary. The bibliography is not a comprehensive one. Entries were chosen primarily for their excellent quality, with exceptions made to achieve representativeness or utility. Only about a fifth of the material read during the search for entries was included. The studies cover biology from the eighteenth century through the mid-twentieth century; other bibliographies in the Garland series cover pre-eighteenth century biological works. All cited items were published prior to 1985 and for the most part were published during the preceding twenty-five years. Exceptions occur for subject areas where there are not many current studies or for a few "classic" works included as examples. One exception was made because there are no current, comprehensive, general histories of biology. Therefore under Biology will be found the old standards and several modern partial surveys, but not the comprehensive work of excellence that one might wish.

Efforts have been made to exclude studies which are medically oriented because this area is so thoroughly indexed and annotated elsewhere. Some popularized works have been included. Most of the works are in English. Selected foreign language books are included; because of my linguistic limitations I relied on book reviews or previous annotations for assistance in describing some of these foreign books.

The full original data base from which this bibliography was compiled included annotated bibliographies such as Smit's <u>History of the Life Sciences</u>, unannotated bibliographies such as <u>Isis Cumulative Bibliography</u>, guides to the literature such as Corsi's, review articles such as Egerton's on ecology. Articles in the journals specific to the field, such as <u>Journal of the History of Biology</u> and <u>Studies in History of Biology</u> were included. This data base was then enlarged by adding citations from monographs and articles as they were read. From this cumulated data base the entries for the bibliography were selected.

The entries include monographs, journal articles, chapters from larger works, edited compilations, and conference proceedings. Collections of "classic" original articles are also included. Bibliographical essay reviews of the historical literature in a specific field of biology are included. Autobiographies and biographies are included only when they focus on the science or a balance between the science and the individual rather than mostly on the individual, and this it seems happened rarely until just recently.

The bibliography is arranged in broad subject areas, then alphabetically by author. The choice of subject arrangement remains problematic because of the shifting disciplinary boundaries in biology. Short of taking the Zen way and placing everything in a single category on the grounds that what is is, decisions had to be made. These were made partly on the basis of Library of Congress subject headings. Interpretation was still necessary in some instances. Thus, heredity, genetics, and molecular biology are seen as aspects of a core concept that changed considerably over time, heredity lasting through the rediscovery of Mendel until T.H. Morgan's <u>Drosophila</u> group when it became genetics which lasted until DNA was discovered when surely it became molecular biology.

The entries include complete bibliographic information, in the <u>Chicago Manual of Style</u> form, and descriptive annotations. Also included are an index to authors, editors, translators, and compilers and an index to subjects.

History of biology, like other areas in the history of science or medicine, has usually been written by the scientists themselves reflecting on their discipline. Their earlier historical research and writing, even at its most able, reflected the previously accepted style of chronological narrative history focused on the great men (persons), great institutions, and great scientific events. These contributions remain important and useful ones even though they do not provide the levels of analysis and interpretation nor the scientific or societal contexts that we now expect in our studies of history biology. Scientists continue to write most history of biology; some of them receive formal training in history as well as in their science. Even those not formally trained in history strive--and some succeed admirably--to reflect in their historical research and writing the currently expected historiography. General historians write history of biology now, too. The published scholarship of these three groups of writers that is included in this bibliography can be characterized as very good indeed.

Because diversity is adaptive, it should be a good sign that in reading the books and journals for this bibliography I encountered not only the several categories of authors mentioned above, but also internalist, externalist, oral, quantitative, revisionist, Whig, intellectual, Kuhnian, social, and Marxist histories---even prosopography. And all of these appear in the bibliography; a novice to the history of biology should experience them all because each opens a different window on the development of biology.

The history of biology is focused at present on evolution, physiology (especially neurophysiology), and heredity/genetics, which mirrors the topics of most interest to biological scientists. Those historical interests continue to be strong ones as is evidenced by the establishment in 1970 of a specialist journal, Journal of the History of Biology. Indeed, the history of biology is flexing its historiographical muscles and those muscles create a vigorous, rigorous history.

I would like to extend my thanks to the many individuals who contributed to the creation of this bibliography. The greatest thanks goes to the staff of the Universitetsbibliotekt I Bergen Avd. Realfagbygget where the majority of my reading took place. Their library does not collect history of biology particularly, but when I found myself there, they willingly borrowed hundreds of books and journals for me through a marvelous Scandinavian on-line computer network. The materials all came, inspiringly, from the great old universities of Northern Europe and they all arrived within two or three days of request. Never have I been so impressed with the willingness and quality of a library staff and its services! A peripatetic bibliographer encounters many libraries. The collection of the Whipple Library at Cambridge was an invaluable resource, and I am grateful for their permission to use it. Here in Minnesota the library sources for the history of biology are in collections (plural) and my thanks goes to librarians in seven libraries and to the Bio-Medical Library staff who patiently requested, received, and returned hundreds of books and journals from those libraries located all over two campuses.

In particular I wish to thank Dr. Michael Osborne for sharing his expertise and for all the excellent advice and encouragement he gave me. To Glenn Brudvig, past Director of the Bio-Medical Library, who always encouraged his staff's scholarly endeavors, I am grateful for support of this project and of my sabbatical leave to work on it. Behind every book lies (1) a great typist---my thanks to Carla Freeman for her after hours efforts, (2) a long-suffering family---my apologies to my husband, daughter, and parents, and (3) a patient and encouraging editor---Ellen Wells deserves extra thanks for her extra patience and encouragement.

Judith A. Overmier
Curator
Owen H. Wangensteen Historical Library
    of Biology and Medicine

REFERENCE

Reference tools for the history of biology are not as plentiful as in other subject areas, such as the health sciences. Good ones do exist however and range from introductory levels, such as this bibliography is, to more sophisticated and specialized approaches. Most novices in a field of historical research come to the library asking for "a book on..." or "everything you've got on..."

BIBLIOGRAPHY

1. Egerton, Frank N. "The History of Ecology: Achievements and Opportunities, Part One." Journal of the History of Biology 16(1983): 259-310.

2. Corsi, Pietro, and Paul Weindling, eds. Information Sources in the History of Science and Medicine. London: Butterworth, 1983.

3. Smit, Piet. History of the Life Sciences: An Annotated Bibliography. Amsterdam: Asher, 1974.

4. Whitrow, Magda, ed. Isis Cumulative Bibliography: A Bibliography of the History of Science Formed From Isis Critical Bibliographies 1-90, 1913-1965. 3 vol. London: Mansell, 1971-1976.

Providing "a book on..." calls for providing the best book, so selective bibliographic tools are required. One of the particular strengths of the history of biology is the bibliographic essay reviews its historians publish. These articles, evaluating the historical literature of a specific topic, help identify the best books and have been included in this bibliography under their appropriate topics. One of Egerton's articles, in which he evaluates histories of nineteenth and twentieth century ecology, is listed here also to provide an example of what to look for. Sometimes these bibliographic essay reviews are part of a larger selective tool such as Corsi and Weindling. Their Information Sources in the History of Science and Medicine introduces the literature, methods, and concepts of the history of science and medicine. It includes two chapters relevant to the history of biology: "Life sciences: natural history" by David Elliston Allen and "Experimentalism and the life sciences since 1800" by Margaret Pelling.

After the category of in-depth evaluations of the literature of the history of biology comes reference tools which identify and briefly annotate the best books, such as this bibliography does. Smit's History of the Life Sciences: An Annotated Bibliography also does this although for a wider subject area. His work annotates a selected bibliography of the history of the life sciences, including medicine. It contains 4,000 entries for books (rarely journal articles) published up to the middle of 1971.

The persons asking for "everything you've got on..." are very well served by the Isis bibliography. Two of Whitrow's volumes contain an invaluable, extensive listing of monographs and journal articles covering all areas of science, arranged by subjects, including biology. This bibliography includes foreign publications as well.

BIOGRAPHICAL DICTIONARIES

GENERAL

5. Gillispie, Charles Coulston, ed. Dictionary of Scientific Biography. 16 vols. New York: Scribner's, 1970-1980.

6. Whitrow, Magda, ed. Isis Cumulative Bibliography: A Bibliography of the History of Science Formed From Isis Critical Bibliographies 1-90, 1913-1965. 3 vol. London: Mansell, 1971-1976.

7. Neu, John, ed. Isis Cumulative Bibliography 1966-1975. A Bibliography of the History of Science Formed from Isis Critical Bibliographies 91-100, Indexing Literature Published from 1965 through 1974. Vol. 1   Personalities and Institutions. London: Mansell, 1980.

It should be remembered that the "on..."in a question can be followed by a person's name as well as by a subject discipline term. Probably the most common of questions is for biographical information. The Dictionary of Scientific Biography is an excellent reference tool which provides original scholarly biographies of deceased significant scientists, biologists included, and identifies the best secondary sources for additional biographical information about them. Whitrow and Neu have compiled volumes for the Isis Cumulative Bibliography that are massive listings of references to biographies of individuals and institutions from the monographic and journal literature. A more useful comprehensive list does not exist.

SPECIFIC

8. Watson, Robert Irving, ed. Eminent Contributors to Psychology. 2 vol. New York: Springer, 1974-1976.

9. Desmond, Ray. Dictionary of British and Irish Botanists and Horticulturists, Including Plant Collectors and Botanical Artists. London: Taylor & Frances, 1977.

10. Fruton, Joseph S. Selected Bibliography of Biographical Data for the History of Biochemistry Since 1800. Philadelphia: American Philosophical Society Library, 1977.

11. Wood, Samuel. Naturalists of the Frontier. University Park, Texas: Southern Methodist University, 1948.

12. Williams, Harley. Great Biologists. London: Bell, 1961.

13. Oliver, F.W., ed. Makers of British Botany: A Collection of Biographies by Living Botanists. Cambridge: Cambridge University Press, 1913.

14. Thomson, J. Arthur. The Great Biologists. London: Methuen, 1932, reprinted 1967.

In addition to the major collective biographies and
bibliographies covering biographies for all of science,
there are numerous smaller specialized reference tools.
Some of these focus on a discipline. Watson's Eminent
Contributors to Psychology, for example, consists of a
bibliography of primary references of 538 noted deceased
psychologists followed by a bibliography of secondary
references. Eminence was determined by a panel of nine
historian psychologists from four countries who
considered 1,040 individuals from the seventeenth to the
twentieth centuries.

Desmond provides bio-bibliographical listings of about
10,000 British and Irish botanists up to 1975. He also
locates their plants, portraits, and manuscripts; he
lists plants named for the entries and gives the name of
the individual who named the plant. Joseph Fruton's
work contains bibliographic references for biographical
material on deceased scientists who participated in the
interaction between biology and chemistry.

Some of the specialized collective biographies have
very narrow coverage. The volume of biographies of ten
naturalists in Texas during 1820-1880 by Samuel Wood is
an example. It is accompanied by a bibliography and an
appendix listing other naturalists known to have worked
in Texas during that period.
It is common for many others of the collective
biographies to cover only the most outstanding
contributors to biology. William's Great Biologists
contains biographies of seven famous biologists:
Aristotle, Linnaeus, Antonius van Leeuwenhoek, Erasmus
Darwin, Charles Darwin, Alfred Russel Wallace, and
Thomas Hunt Morgan. Oliver has compiled 17 biographies
of famous British botanists and 1 group biography of
Edinburgh botany professors, 10 of which were written by
botanists originally as lectures, six added for
publication. Thompson's The Great Biologists contains
biographical presentations of the life and contributions
of 27 biologists from Aristotle to Ray Lankester. These
small bibliographies, bio-bibliographies and collective
biographies are very common and many others than these
few examples will be found in comprehensive listings,
such as Isis.

DICTIONARY OF THE HISTORY OF SCIENCE

15. Bynum, W.F., E.J. Browne, Roy Porter, eds. Dictionary of the History of Science. London: Macmillan, 1981.

It is fortunate for the novice in the history of biology that there is a dictionary covering the history of science. It contains descriptive definitions of fields of science and of important terms and concepts related to them. Entries include cross references and often bibliographies also. A name index at the end provides dates, identification of subject fields and research interests of scientists. Represents biology very well and is an essential reference work for the historian of biology.

THE HISTORY OF BIOLOGY

ANATOMY

16. Billings, Susan M. "Concepts of Nerve Fiber
    Development, 1839-1930." Journal of the History
    of Biology 4(1971): 275-305.

    Discusses three theories of how the nerve fiber is
    formed that were debated between Theodor Schwann's ideas
    in 1839 and 1930 when the outgrowth theory was generally
    agreed upon. Describes the observational and
    experimental studies necessary for the understanding of
    nerve fiber development.

17. Choulant, Ludwig. History and Bibliography of
    Anatomic Illustration. Translated and annotated
    by Mortimer Frank. New York: Hafner, 1962,
    c1945.

    Remains the standard source for information about
    and discussion of anatomical illustration from antiquity
    through the nineteenth century. Provides brief
    biographical information about the anatomists and
    discusses the illustrations in their anatomical works.

18. Meyer, Alfred. Historical Aspects of Cerebral
    Anatomy. London: Oxford University Press, 1971.

    Contains essays on the history of cerebral anatomy
    into the twentieth century arranged in three regions:
    the basal ganglia and the diencephalon, the olfactory
    system, the cerebral convolutions and fissures.

19. Rook, Arthur, ed. Cambridge and Its Contribution
    to Medicine. (Proceedings of the Seventh British
    Congress on the History of Medicine, University
    of Cambridge, 10-13 September, 1969.) London:
    Wellcome Institute of the History of Medicine,
    1971.

    Includes several minor presentations on the
    history of anatomy, physiology, biochemistry, and
    genetics at Cambridge University.

20. Weindling, Paul. "Theories of the Cell State in
     Imperial Germany." In Biology, Medicine and
     Society, 1840-1940, edited by Charles Webster,
     99-155. Cambridge: Cambridge University Press,
     1981.

     Proposes an analogy between the cell as elementary
organism and as a social and political concept. Focuses
on the social and institutional factors influencing
Berlin anatomy from the 1880s to 1918, specifically
through the work of Oscar Hertwig and Wilhelm Waldeyer.
Describes their scientific use of social concepts.

ANIMAL RESEARCH

21. French, Richard D. Antivivisection and Medical
     Science in Victorian Society. Princeton:
     Princeton University Press, 1975.

     Includes discussion of the slow development of
experimental physiology in England and the problems
physiologists had in carrying out their research.

22. Rowan, Andrew N., and Bernard E. Rollin. "Animal
     Research - For and Against: A Philosophical,
     Social, and Historical Perspective."
     Perspectives in Biology and Medicine 27(1983): 1-
     17.

     Outlines briefly in the first half of the article
the history of opposition to animal research, especially
in America.

23. Schiller, Joseph. "Claude Bernard and
     Vivisection." Journal of the History of Medicine
     22(1967): 246-247.

     Discusses the importance Claude Bernard placed on
vivisection as one of the constituents of the
experimental method; mentions also the role of
vivisection in the development of physiology as an
independent experimental science.

24. Schiller, Joseph. Claude Bernard et les Problèmes
    Scientifiques de Son Temps. Paris: Les Editions
    de Cèdre, 1967.

    Discusses thoughtfully Claude Bernard in the
context of the scientific issues of his time, including
a chapter on vivisection.

25. Sechzer, Jeri A. "Historical Issues Concerning
    Animal Experimentation in the United States."
    Social Science and Medicine [F] 15F(1981): 13-17.

    Compares very briefly the development of movements
against animal research in Great Britain and the United
States.

26. Turner, James. Reckoning with the Beast: Animals,
    Pain, and Humanity in the Victorian Mind.
    Baltimore: Johns Hopkins University Press, 1980.
    (Johns Hopkins University. Studies in Historical
    and Political Science, 98th ser. no. 2)

    Includes discussion of the Victorians' emotional
response to the use of animals in the newly developing
experimental physiology and medicine.

BIOCHEMISTRY

27. Aulie, Richard C. "Boussingault and the Nitrogen
    Cycle." Proceedings of the American
    Philosophical Society 114(1970): 435-479.

    Describes the experiments on the nitrogen cycle
published from 1836 to 1876 by Jean-Baptiste
Boussingault.

28. Debru, Claude. L'Esprit des Proteines: Histoire et
    Philosophie Biochimiques. Paris: Hermann, 1983.

    Contains a survey of the history of biochemistry
in general to the end of the nineteenth century and then
focuses in research on the proteins.

6   Biochemistry

29.   Edsall, John T. "Blood and Hemoglobin: The
      Evolution of Knowledge of Functional Adaptation
      in a Biochemical System." Journal of the History
      of Biology 5(1972): 205-257.

      Explains clearly the research through 1930 on the
role of the circulating blood in the respiratory system
that established its biochemical basis. Focuses on the
hemoglobin molecule and its role in the transport of
oxygen and carbon dioxide.

30.   Fruton, Joseph S. Molecules and Life; Historical
      Essays on the Interplay of Chemistry and Biology.
      New York: Wiley-Interscience, 1972.

      Surveys five areas of biochemistry from 1800-
1950. Discusses research on ferments, enzymes,
proteins, nuclein to the double helix, intracellular
respiration, and biochemical pathways in metabolism.

31.   Geison, Gerald L. "The Protoplasmic Theory of Life
      and the Vitalist-Mechanist Debate." Isis
      60(1969): 273-292.

      Describes Thomas Henry Huxley and Lionel S.
Beale's conflicting concepts about protoplasm.

32.   Glas, Eduard. Chemistry and Physiology in Their
      Historical and Philosophical Relations. Delft,
      The Netherlands: Delft University Press, 1979.

      Surveys the growth of physiological chemistry from
1770 to 1890, with a final chapter on the emergence of
modern biochemistry after 1890.

33.   Goodwin, T.W., ed. British Biochemistry Past and
      Present. London: Academic Press, 1970.

      Contains thirteen papers with varying levels of
historical treatment written by major participants in
the fields. Covers 1911 through 1964; arranged in four
broad areas - molecular biology, immunology,
intermediary metabolism, and separation methods.

34. Grmek, Mirko Drazen. "First Steps in Claude
     Bernard's Discovery of the Glycogenic Function of
     the Liver." Journal of the History of Biology
     1(1968): 141-154.

     Analyzes Claude Bernard's laboratory journals
which document and detail numerous experiments on the
destination of sugar in animal organisms beginning in
1843 and continuing to 1848 when an unexpected
experimental result changed the course of his research.

35. Kohler, Robert E. "The Background to Eduard
     Buchner's Discovery of Cell-Free Fermentation."
     Journal of the History of Biology 4(1971): 35-61.

     Discusses controversy over alcoholic fermentation,
Hans and Eduard Buchner's researches, Eduard Buchner's
discovery of cell-free fermentation, and the importance
of the discovery of zymase to the development of the new
field of biochemistry.

36. Kohler, Robert E. "The Background to Otto
     Warburg's Conception of the Atmungsferment."
     Journal of the History of Biology 6(1973): 171-
     192.

     Describes the Atmungsferment, Otto Warburg's
theory of respiration, and his research on the topic
from 1908 to 1925.

37. Kohler, Robert E. "The Enzyme Theory and the
     Origin of Biochemistry." Isis 63(1974): 181-196.

     Discusses the background of the development of
biochemistry as a profession after the turn of the
century. Identifies features that distinguish it from
the earlier physiological chemistry.

38. Kohler, Robert E. "The History of Biochemistry: A
     Survey." Journal of the History of Biology
     8(1975): 275-318.

     Contains evaluative reviews of recent (five years)
published work on the history of biochemistry.
Discusses historiography of biochemistry including

themes and problems of the field and historiographic
approaches to the field.

39. Kohler, Robert E.   "The Reception of Eduard
     Buchner's Discovery of Cell-Free Fermentation."
     Journal of the History of Biology 5(1972): 327-
     353.

     Begins with Eduard Buchner's announcement of the
1897 discovery of cell-free fermentation and his
identification of zymase as an enzyme. Analyzes the
challenge this presented to the protoplasm theory, the
ensuing debate and its resolution, and explains its role
in the advance of the developing field of biochemistry.

40. Kohler, Robert E.   "Rudolf Schoenheimer, Isotopic
     Tracers, and Biochemistry in the 1930's."
     Historical Studies in the Physical Sciences
     8(1977): 257-298.

     Describes Rudolf Schoenheimer's interest in
organic chemistry, his identification of intermediary
metabolism as a major research problem, his use of
isotopes in its study, his theory of the dynamic state
of body constituents, and his research school. Treats
these factors as exemplars of the development of
biochemistry in the 1930s.

41. Kottler, Dorian B.   "Louis Pasteur and Molecular
     Dissymmetry, 1844-1857." Studies in History of
     Biology 2(1978): 57-98.

     Covers Pasteur's transition from work on inorganic
to organic substances. Analyzes in detail Pasteur's
crystallographic research from 1844 to his research on
fermentation. Identifies his connection between
molecular dissymmetry and life as a critical turning
point.

42. Krebs, Hans, and Rosewitha Schmid.   Otto Warburg:
     Cell Physiologist, Biochemist, and Eccentric.
     Translated by Hans Krebs and Anne Martin.
     Oxford: Clarendon, 1981.

Outlines Otto Warburg's main scientific achievements.

43. Leicester, Henry M. Development of Biochemical Concepts from Ancient to Modern Times. Cambridge: Harvard University Press, 1974.

Discusses biochemical concepts from ancient times through the 1930s, including digestion and assimilation, enzymes, vitamins, and hormones.

44. Lieben, Fritz. Geschichte der Physiologischen Chemie. New York: Georg Olms, 1970.

Reprint of the 1935 standard survey of the history of biochemistry from the organic chemistry viewpoint.

45. Morgan, Neil. "The Development of Biochemistry in England through Botany and the Brewing Industry (1870-1890)." History and Philosophy of the Life Sciences 2(1980): 141-166.

Describes British botany in the 1860s and 1870s and its transformation by the introduction of German physiological botany concepts. Explains the brewing industry and discusses how it and plant physiologists contributed to the knowledge of plant enzymes.

46. Needham, Dorothy. Machina Carnis: The Biochemistry of Muscular Contraction in Its Historical Development. Cambridge: Cambridge University Press, 1971.

Details comprehensively theories of and research on muscle excitation and contraction from the earliest times into the 1960s.

47. Needham, Joseph, ed. The Chemistry of Life: Eight Lectures on the History of Biochemistry. Cambridge: Cambridge University Press, 1970.

Contains eight historical essays by seven Cambridge biochemists and Mikulás Teich, historian of biochemistry, on the history of biochemistry primarily after 1800 and covering photosynthesis, enzymes, microbiology, neurology, hormones, vitamins, and nineteenth century and modern biochemistry.

48.   Needham, Joseph, ed. Hopkins and Biochemistry, 1861-1947. Cambridge: Heffer, 1949.

Contains papers commemorating Sir Frederick Gowland Hopkins' teaching, scientific, and personal influence. Includes fifteen of Hopkins' addresses and a bibliography of his works.

49.   Teich, Mikulás. "Ferment or Enzyme: What's In A Name?" History and Philosophy of the Life Sciences 3(1981): 193-215.

Outlines research on the chemistry and biology of fermentation from Antoine Laurent Lavoisier through Eduard Buchner. Suggests that the knowledge that chemical changes in the cell could not be identified with fermentation brought about the use of the term enzyme instead of ferment.

50.   Teich, Mikulás. "From 'Enchyme' to 'Cytoskeleton': The Development of Ideas on the Chemical Organization of Living Matter." In Changing Perspectives in the History of Science, edited by Mikulás Teich and Robert Maxwell Young, 439-471. London: Heinemann, 1973.

Traces biological and chemical research contributions to cell chemistry and cell morphology from about 1840 to 1940; highlights the importance of linking the two fields.

51.   Teich, Mikulás. "On the Historical Foundations of Modern Biochemistry." Clio Medica 1(1965): 41-57.

Traces the evolving nineteenth century relationships between chemistry and organic chemistry, organic chemistry and physiology, and physiology and biochemistry.

52. Young, F.G. "The Rise of Biochemistry in the
     Nineteenth Century, with Particular Reference to
     the University of Cambridge." In <u>Cambridge and
     Its Contribution to Medicine</u>, edited by Arthur
     Rook, 155-172. London: Wellcome Institute of the
     History of Medicine, 1971.

     Contrasts the development of British biochemistry
with continental. Suggests reasons for its slowness.
Discusses the growth of biochemistry at Cambridge
through the work of Michael Foster, Sheridan Lea, and
Gowland Hopkins.

BIOGEOGRAPHY

53. Browne, Janet. <u>The Secular Ark: Studies in the
     History of Biogeography</u>. New Haven: Yale
     University Press, 1983.

     Examines the development of biogeography through
the theories of animal and plant geographical
distribution studied by geologists, geographers,
botanists, zoologists, paleontologists, and natural
historians. Discusses two main threads of thought, that
of geologists and paleontologists interested in the
history of distribution and of botanists and zoologists
interested in patterns. Concludes with the work of
Darwin and Wallace when the research from the wide
variety of fields began to come together.

54. Fichman, Martin. "Wallace, Zoogeography and the
     Problem of Landbridges." <u>Journal of the History
     of Biology</u> 10(1977): 45-63.

     Discusses the development of Alfred Russel
Wallace's concepts of zoogeography from his original
belief in land bridges to his mature work as it appeared
in his <u>Geographical Distribution of Animals</u> and <u>Island
Life</u>.

55. Kinch, Michael Paul. "Geographical Distribution and
     the Origin of Life: The Development of Early
     Nineteenth-Century British Explanations."
     <u>Journal of the History of Biology</u> 13(1980): 91-
     119.

Presents nineteenth century pre-Darwinian theories of the origin and distribution of life proposed by naturalists, in particular by James Cowles Prichard, William Swainson, William Kerby, Philip Luttey Sclater, and Charles Lyell.

56. Nelson, Gareth. "From Candolle to Croizat: Comments on the History of Biogeography." Journal of the History of Biology 11(1978): 269-305.

Traces ideas of biogeography back to 1761 and Buffon's law and discusses approaches to these ideas up through Leon Croizat, with emphasis on the contributions of A.P. de Candolle.

57. Sterling, Keir B., ed. Selections from the Literature of American Biogeography. New York: Arno, 1974.

Contains reprints of twenty-eight landmark contributions to biogeography during the period 1830-1948; accompanied by brief introduction.

BIOLOGY

58. Allen, Garland E. Life Science in the Twentieth Century. Cambridge: Cambridge University Press, 1978.

Describes the changes in biology from the late nineteenth century to mid-twentieth century in terms of its reorientation from descriptive and speculative natural history and mechanistic physiology to experimental biology with emphasis on the molecular level, and finally to a synthesis. Analyzes the areas of embryonic development, heredity, evolution, general physiology, biochemistry, and molecular biology as exemplars.

59. Arthur F.W. The American Biologist through Four Centuries. Springfield: Thomas, 1982.

Surveys at an introductory level four centuries

of the life sciences in America, including
practitioners, institutions, and the various
disciplines.

60. Asimov, Isaac. A Short History of Biology.
     Westport: Greenwood, 1980.

     Provides a short, readable overview of modern
biology written at a very basic level.

61. Bodenheimer, F.S. The History of Biology: An
     Introduction. London: William Dawson & Sons,
     1958.

     Contains an introductory level discussion of
historiography of science, a short history of biology,
and excerpts from classics in the history of science.

62. Cahn, Théophile. La Vie et l'Oeuvre d'Etienne
     Geoffroy Saint Hilaire. Paris: Presses
     Universitaires de France, 1962.

     Discusses the life and work of Etienne Geoffroy
Saint-Hilaire.

63. Canguilhem, Georges. Études d'Histoire et de
     Philosophie des Sciences. 2d ed. Paris:
     Librarie Philosophique J. Vrin, 1970.

     Contains reprints of twenty-five papers, many on
biological topics such as Claude Bernard's experimental
method or the concept of the reflex in the nineteenth
century.

64. Caullery, Maurice. A History of Biology.
     Translated by James Walling. New York: Walker,
     1966.

     Contains a very brief survey at an introductory
level of a few major contributions to biology from the
Greeks to the present.

65. Coleman, William. Biology in the Nineteenth
     Century: Problems of Form, Function, and
     Transformation. Cambridge: Cambridge University
     Press, 1977.

     Discusses aspects of nineteenth century biology,
in particular the subjects cell theory, embryology,
evolution, physical anthropology, human paleontology,
and physiology under the three themes of form, function,
and transformation. Covers historical explanation in
biology and its displacement by the rise of experimental
investigation.

66. Dawes, Ben. A Hundred Years of Biology. London:
     Duckworth, 1952.

     Outlines the main trends in biology, focusing on
early twentieth century. Includes agricultural and
marine biology, all animal behavior and a chapter on
technical advances along with the more common topics of
taxonomy, evolution, reproduction, development, etc.

67. Gabriel, Mordecai L., and Seymour Fogel. Great
     Experiments in Biology. Englewood Cliffs:
     Prentice-Hall, 1955.

     Reprints classic experiments from cell theory,
general physiology, microbiology, plant physiology,
embryology, genetics, and evolution. Each field is
accompanied by a chronology.

68. Gardner, Eldon J. History of Biology. 3d ed.
     Minneapolis: Burgess, 1972.

     Covers the history of biology from ancient Greece
through modern genetics at the introductory textbook
level with emphasis on the contributions of outstanding
scientists.

69. Lanham, U. Origins of Modern Biology. New York:
     Columbia University Press, 1968.

     Contains five minor chapters (of eleven) written
in a popular style about biology since the 1700s.

70. Lenoir, Timothy. "The Göttingen School and the
     Development of Transcendental Naturphilosophie in
     the Romantic Era." Studies in the History of
     Biology 5(1981): 111-205.

     Explores the development of the Göttingen School
of Biology during the latter half of the eighteenth
century. Focuses on the ideas of Johann Friedrich
Blumenbach and several of his students, concluding with
Gottfried Reinhold Treviranus and the publication of his
Biologie.

71. Locy, William A. Biology and Its Makers. New
     York: Henry Holt, 1908.

     Outlines the history of biology in a popular style
from the biographical perspective. Covers pre-
evolutionary biology (sans botany) in the first fifteen
chapters; ends with five chapters on evolution.

72. McCullough, Dennis M. "W.K. Brooks' Role in the
     History of American Biology." Journal of the
     History of Biology 2(1969): 411-438.

     Reevaluates the influence of William Keith Brooks
on the development of the biology of the major
biologists such as E.B. Wilson, T.H. Morgan, R.C.
Harrison, E.G. Conklin and W. Bateson, who were his
students. Assigns more weight to graduate education at
Johns Hopkins University, to the introduction of
European experimentalism, and to working at Woods Hole.

73. Magner, Lois N. A History of the Life Sciences.
     New York: Marcel Dekker, 1979.

     Contains chapters on several aspects of the life
sciences, including the microscope, generation,
reproduction, and development, cell theory, microbiology
and biogenesis, physiology, evolution, genetics and
molecular biology. Written in a popular style.

74. Mendelsohn, Everett. "The Biological Sciences in
     the Nineteenth Century: Some Problems and
     Sources." History of Science 3(1964): 39-59.

Reviews historical treatments of nineteenth century biology and suggests new areas, areas needing further work, and new questions for the historian to approach.

75.   Nordenskiold, Erik.   History of Biology: A Survey. Translated from the Swedish by Leonard Bucknall Eyre.   New York: Knopf, 1928; New York: Tudor, 1949.

     Remains the standard source for the general history of biology.   Covers antiquity through the nineteenth century, with accounts of anatomy, embryology, natural history, physiology, and evolution.

76.   Ospovat, Dov.   "Perfect Adaptation and Teleological Explanation: Approaches to the Problem of Life in the Mid-Nineteenth Century."   Studies in History of Biology 2(1978): 33-56.

     Discusses changes in mid-nineteenth century biological explanation.   Proposes a teleological-non-teleological dichotomy as an appropriate approach to understanding the views of mid-nineteenth century naturalists.

77.   Pauly, Philip J.   "The Appearance of Academic Biology in Late Nineteenth-Century America." Journal of the History of Biology 17(1984): 369-397.

     Describes efforts to establish biology as a discipline at six American universities from 1870 to 1900.   Emphasizes the role of the development of scientific medicine in the discussion of reasons for their success or failure.

78.   Radl, Emanuel.   History of Biological Theories. Translated and adapted by E.J. Hatfield.   London: Oxford University Press, 1930.

     Covers the changes in biology from Darwin through Driesch from an anti-Darwin perspective.   Translation and adaptation of only a portion of Geschichte der biologischen Theorie, 2 vols., 1905-1909.

79. Singer, Charles. A History of Biology to about the Year 1900: A General Introduction to the Study of Living Things. 3d and revised ed. London and New York: Abelard-Schuman, 1959.

Surveys biology from ancient times to early twentieth century at the introductory level. Treats the main themes of modern biology in seven chapters - "Cell and Organism," "Essentials of Vital Activity," "Relativity of Functions," "Biogenesis and its Implications," "Development of the Individual," "Sex," and "Mechanisms of Heredity."

80. Sirks, M.J., and Conway Zirkle. The Evolution of Biology. New York: Ronald, 1964.

Surveys the history of biology from ancient to modern times, arranged around developmental phases such as specialization.

81. Thomson, J. Arthur. The Science of Life. An Outline of the History of Biology and Its Recent Advances. London: Blackie, 1899.

Surveys broadly at an introductory level the history of biology up to the turn of the century, with chapters on such topics as psychology of animals, bionomics, geographical distribution, paleontology, heredity, animal morphology, and plant morphology.

BOTANY

82. Allan, Mea. Darwin and His Flowers: The Key to Natural Selection. London: Farber & Farber, 1977.

Focuses on Darwin's botanical contributions bringing together, in a popular style accompanied by numerous illustrations, information about his plant collections from the Beagle voyage, his botanically related correspondence and notes, his published work on plants, and his botanical research.

83.   Berkeley, Edmund. "The History of the Naming of
      Loblolly Bay." Journal of the History of Biology
      3(1970): 149-154.

         Describes the fifteen year effort of Dr. Alexander
      Garden of South Carolina to have the Loblolly Bay named.

84.   Daudin, Henri. Études d'Histoire des Sciences
      Naturelles. 2 vols. Paris: Alcan, 1926.

         Remains a classic study in two volumes - the
      first, De Linné à Jussieu: Méthodes de la Classification
      et Idée de Série en Botanique et en Zoologie (1740-1790)
      and the second, Cuvier et Lamarck: Les Classes
      Zoologiques et Idée de Série Animale (1740-1790).

85.   Dupree, A. Hunter. Asa Gray, 1810-1888.
      Cambridge: Harvard University Press, 1959.

         Details Asa Gray's professional career, his
      international role, and his relationship with Louis
      Agassiz.

86.   Eriksson, Gunnar. "Linnaeus the Botanist." In
      Linnaeus, The Man and His Work, edited by Tore
      Frangsmyr, 63-109. Berkeley: University of
      California Press, 1983.

         Details the influences on and the development of
      Linnaeus' classification by sexual characters and
      discusses others of his biological ideas.

87.   Ewan, Joseph, ed. A Short History of Botany in
      the United States. New York: Hafner, 1969.

         Collects brief essays by authorities in the fields
      on the history of thirteen areas of plant science
      including plant morphology, anatomy, genetics, cytology,
      physiology and specialties such as pteridoly, bryology
      and lichenology.

88.   Green, J. Reynolds. A History of Botany 1860-1900.
      Being a Continuation of Sachs `History of Botany,

1530-1860.' Oxford: Clarendon Press, 1909.
Remains a classic survey of botanical history
arranged in three parts: morphology, anatomy, and
physiology (the latter comprising half the book).

89. Hagen, Joel B. "Experimentalists and Naturalists
     in Twentieth-Century Botany: Experimental
     Taxonomy, 1920-1950." Journal of the History of
     Biology 17(1984): 249-270.

Provides examples of botanists combining
traditional and experimental taxonomy and of cooperative
research among biologists in the fields of taxonomy,
cytology, ecology, and genetics. Discusses the research
programs of F.E. Clements, H. M. Hall, Gote Turesson,
E.B. Babcock, David Keck, William Hiesey, and Jens
Clausen.

90. Montgomery, William M. "The Origins of the Spiral
     Theory of Phyllotaxis." Journal of the History
     of Biology 3(1970): 299-323.

Presents the new theory of Carl Schimper and
Alexander Braun on the distribution of leaves around an
axis.

91. Morton, A.G. History of Botanical Science: An
     Account of the Development of Botany from Ancient
     Times to the Present Day. London: Academic
     Press, 1981.

Surveys the complete field of botany at an
introductory level through the nineteenth century.

92. Overfield, Richard A. "Charles E. Bessey: The
     Impact of the 'New' Botany on American
     Agriculture, 1880-1910." Technology and Culture
     16(1975): 162-181.

Delineates the significant role of Charles E.
Bessey in effecting American botany's change from
systematics to an evolutionary, experimental science
that might encompass physiology, pathology, ecology, and
plant geography. Describes his contributions to the

professionalization of American botany and to botany's impact on agriculture.

93. Reed, Howard Sprague. A Short History of the Plant Sciences. Waltham, Mass.: Chronica Bot., 1942.

Surveys the plant sciences chronologically from the earliest times and topically (plant geography, morphology, cytology, etc.) at an introductory level using the biographical approach; an old standard.

94. Rodgers, Andrew Denny. American Botany 1873-1892: Decades of Transition. Princeton: Princeton University Press, 1944.

Covers the field broadly, discusses great botanists, surveys and explorations, development of morphology and physiology, botanical laboratories, paleobotany and ecology; an old standard.

95. Sachs, Julius von. A History of Botany. Translated by Henry E.F. Garnsey; revised by Isaac Bayley Balfour. Oxford: Clarendon Press, 1906.

Covers 1530-1860 in three books: history of morphology and classification, history of vegetable anatomy, and history of vegetable physiology.

96. Stafleu, Frans A. Linnaeus and the Linnaeans. The Spreading of Their Ideas in Systematic Botany, 1735-1789. Utrecht: Oosthoek, 1971.

Explores the development of Linnaeus' ideas in systematic botany and their reception in the Netherlands, Austria, Great Britain, Germany, Switzerland, and France from 1735 to 1789. Includes a brief chapter on Linnaeus' students at Uppsala and their 186 dissertations.

97. Stevens, P.F. "Hauy and A.P. Candolle: Crystallography, Botanical Systematics, and Comparative Morphology, 1780-1840." Journal of the History of Biology 17(1984): 49-82.

Focuses on botanist Augustin-Pyramus de Candolle and his natural system. Discusses the relationship between botanical systems and crystallography.

98. Von Maltzen, Kraft E. "New Formation of Organs in Plants - The Foundation of Plant Morphogenesis." Journal of the History of Biology 4(1971): 307-317.

Covers the nineteenth and twentieth century research of Johannes von Hanstein, Hermann Vochting, Julius Sachs, George Klebs, Karl Goebel, and Gottlieb Haberlandt on the site of new formation of organs in plants.

99. Weevers, T. Fifty Years of Plant Physiology. Amsterdam: Scheltema and Holkema, 1949.

Remains a classic treatment of plant physiology from an historical perspective. Covers the period 1895 to 1945, the period following the histories of Sachs and Green.

100. Zirkle, Conway. The Beginnings of Plant Hybridization. Philadelphia: University of Pennsylvania, 1935.

Describes plant breeding, focusing on the period before Koelreuter, particularly on the eighteenth century.

CYTOLOGY

101. Baker, John R. "The Cell Theory: A Restatement, History and Critique." Quarterly Journal of Microscopical Science 89, 90, 93, 94, 96(1948-1955): 103-125, 87-108, 157-190, 407-440, 449-481.

Defends the cell-theory in a series of explanatory historical articles organized around seven propositions which restate the cell-theory.

102. Baltzer, Fritz. Theodor Boveri: Life and Work of a Great Biologist. Translated by Dorothea Rudnick. Berkeley: University of California Press, 1967.

   Discusses the life and scientific contributions of the nineteenth century cytologist Theodor Boveri, including his chromosomal and his embryological research.

103. Hughes, Arthur. A History of Cytology. London and New York: Abelard-Schuman, 1959.

   Surveys cytology from the development of the microscope and the discovery of cells.

104. Rather, L.J. Addison and the White Corpuscles: An Aspect of Nineteenth-Century Biology. London: Wellcome Institute of the History of Medicine, 1972.

   Describes William Addison's research on white blood cells in detail and in the context of nineteenth century biology.

DEVELOPMENT

105. Oppenheimer, Jane M. "The Growth and Development of Developmental Biology." In Major Problems in Developmental Biology, edited by Michael Locke, 1-27. New York: Academic Press, 1966.

   Surveys the field of developmental biology focusing on 1939 when the first symposium on development and growth was held and the following two decades.

106. Temkin, Owsei. "German Concepts of Ontogeny and History around 1800." Bulletin of the History of Medicine 24(1950): 227-246.

   Describes the concept of stages of life held by German physiologists around 1800. Discusses the roles of history and embryology, particularly epigenesis, on the development of this concept.

ECOLOGY

107.   Allee, Warder Clyde, and T. Park. "The History of
        Ecology." In Principles of Animal Economy,
        edited by Warder Clyde Allee et al., 13-72.
        Philadelphia: Saunders, 1950.

        Outlines the pre-1900 background and growth of
ecology and then lists developments of various aspects
of the field during the first four decades of the
twentieth century.

108.   Andrewartha, H.G., and L.C. Birch. "The History
        of Insect Ecology." In History of Entomology,
        edited by Ray F. Smith, Thomas E. Mittler, and
        Carroll N. Smith, 229-266. Palo Alto: Annual
        Reviews, 1973.

        Describes the ecological work of entomologists,
emphasizing population ecology as the main area of
interest for entomology.

109.   Chisholm, Anne. Philosophers of the Earth:
        Conversations with Ecologists. New York:
        Dutton, 1972.

        Contains short, popularized accounts of
interviews with eighteen major contributors to modern
ecology, focusing on its relationships with
conservation.

110.   Cittadino, Eugene. "Ecology and the
        Professionalization of Botany in America, 1890-
        1905." Studies in the History of Biology
        4(1980): 171-198.

        Discusses the early period of American ecology
(1890-1905) as one aspect of American botany's turning
from description and classification to the study of
process and function. Emphasizes the work of lesser
known contemporaries of Frederic E. Clements and Henry
Chandler Cowles (who dominated the field after 1905) and
includes such botanists as Frederick V. Coville, Thomas
H. Kearney, Daniel T. MacDougal, Conway MacMillan, A.S.

Hitchcock, Volney M. Spalding, and William Francis
Ganong.

111.   Egerton, Frank N., ed. American Plant Ecology,
       1897-1917. New York: Arno, 1977.

       Contains reprints of thirteen classic papers on
plant ecology.

112.   Egerton, Frank N. "A Bibliographical Guide to the
       History of General Ecology and Population
       Ecology." History of Science 15(1977): 189-215.

       Provides discussion of publications on the
history of ecology under the subheadings general
ecology, population ecology, competition, and balance of
nature; a good source.

113.   Egerton, Frank N. Early Marine Ecology. New
       York: Arno, 1977.

       Contains reprints of six nineteenth century and
one early twentieth century contributions to marine
ecology by six noted scientists including Edward Forbes,
Ernst Haeckel, and Victor Hensen.

114.   Egerton, Frank N. "Ecological Studies and
       Observations Before 1900." In Issues and Ideas
       in America, edited by Benjamin J. Taylor and
       Thurman J. White, 311-351. Norman: University
       of Oklahoma Press, 1976.

       Describes mid-nineteenth century American ecology
as observational, descriptive, and classificatory.
Identifies four areas of late nineteenth century
developing ecological science: oceanography, limnology,
plant ecology, and animal ecology.

115.   Egerton, Frank N. History of American Ecology.
       New York: Arno, 1977.

       Brings together seven reprints and two new
articles on the history of American ecology. Focuses on

the twentieth century. Includes the classic work by
David G. Frey on the Birge-Juday limnology era in
Wisconsin and a history of the Ecological Society of
America by Robert L. Burgess.

116. Egerton, Frank N. "The History of Ecology:
      Achievements and Opportunities, Part One."
      Journal of the History of Biology 16(1983): 259-
      310.

      Reviews histories of nineteenth and twentieth
century ecology; excellent source.

117. Egerton, Frank N. "Richard Bradley's
      Understanding of Biological Productivity."
      Journal of the History of Biology 2(1969): 391-
      410.

      Examines Richard Bradley's discussion of
agricultural productivity in terms of investment and
profit and his application of this to trees, grapevines,
rabbits, and fish, among others, in eighteenth century
England.

118. Egerton, Frank N. "Studies of Animal Populations
      from Lamarck to Darwin." Journal of the History
      of Biology 1(1968): 225-259.

      Discusses first the early nineteenth century
ideas of Georges Cuvier, John Playfair, Augustin-Pyramus
de Candolle, Charles Lyell, and others, about population
pressure, competition, extinction, biogeography, and
biological communities. Closes with a discussion of
Darwin's use of these ideas.

119. Kingsland, Sharon. "The Refractory Model: The
      Logistic Curve and the History of Population
      Ecology." Quarterly Review of Biology 57(1982):
      29-52.

      Describes Raymond Pearl's introduction of the
logistic curve as a law of growth, its repudiation in
that form and its eventual acceptance as a demographic
tool through the personal efforts of Pearl and the
support of Alfred James Lotka and Georgii Frantsevich.

120.   Kormondy, Edward J., and J. Frank McCormick.
       Handbook of Contemporary Developments in World
       Ecology. Westport: Greenwood, 1981.

       Includes short historical summaries of ecology in
specific countries in chapters discussing mainly
contemporary ecology in those countries.

121.   Lowe, P.D.   "Amateurs and Professionals: The
       Institutional Emergence of British Plant
       Ecology."   Society of the Bibliography of
       Natural History Journal 7(1976): 517-535.

       Discusses the professionalization of plant
ecology in nineteenth century Great Britain; emphasizes
the local and regional natural history societies with
both amateur and scientist memberships, and the roles
played by both as ecology became more scientific and
specialized.

122.   Lussenhop, John.   "Victor Hensen and the
       Development of Sampling Methods in Ecology."
       Journal of the History of Biology 7(1974): 319-
       337.

       Recounts Victor Hensen's development of
statistical methods to sample fish eggs and plankton.
Explains the difficulties in acceptance of them that he
encountered.

123.   McIntosh, Robert P.   "Ecology Since 1900."   In
       Issues and Ideas in America, edited by Benjamin
       J. Taylor and Thurman J. White, 353-372.
       Norman: University of Oklahoma Press, 1976.

       Divides twentieth century American ecology into
three periods: 1900-1920 in which McIntosh describes the
establishment of a professional society and journal and
the beginnings of specialization; 1920-1950 in which he
identifies the scientific concerns of ecology as it
introduces quantification and develops working
principles; and post World War II in which he notes
ecology's transition to a "big science," its
increasingly mathematical approach (especially modeling)
and its development of a theoretical base.

124.   Phytopathological Classics of the Nineteenth
        Century. New York: Arno, 1977.

        Provides reprints and/or translations of ten
classic selections by Benedict Prevost, Agostino Bassi,
Miles Joseph Berkeley, Michael Stephanovitch Woronin,
Pierre Marie Alexis Millardet, Adolf Mayer, Dmitrii
Ivanowski, Martinus W. Beijerinck, and Erwin Baur. Each
is accompanied by a brief biography.

125.   Scudo, Francesco M., and James R. Ziegler, eds.
        The Golden Age of Theoretical Ecology: 1923-
        1940. A Collection of Works by Volterra,
        Koslitzin, Lotka, and Kolmogoroff. Berlin:
        Springer-Verlag, 1978.

        Contains translations of twenty-three classic
papers on ecology, arranged in five sections: "Logistic
Approaches;" "Competition and Predation;" "Parasitism,
Epidemics and Symbiosis;" "Genotypic Selection and
Evolution;" and "Life and the Earth," with each section
accompanied by a short introduction.

126.   Stauffer, Robert Clinton. "Ecology in the Long
        Manuscript Version of Darwin's Origin of the
        Species and Linnaeus' Oeconomy of Nature."
        Proceedings of the American Philosophical
        Society 104(1960): 235-241.

        Identifies ecological viewpoints in Darwin's
published and unpublished work; traces some of the roots
of these views to Charles Lyell and Linnaeus.

127.   Tobey, Ronald C. "American Grassland Ecology,
        1895-1955: The Life Cycle of a Professional
        Research Community." In History of American
        Ecology, edited by Frank N. Egerton. New York:
        Arno Press, 1977.

        Utilizes the Crane-Price/Crane-Kuhn model to
analyze the development of the study of grassland
ecology in America from 1895 to 1955.

128.   Tobey, Ronald C. Saving the Prairies: The Life
        Cycle of the Founding School of American Plant

Ecology, 1895-1955. Berkeley: University of
California Press, 1981.

Discusses Charles Edwin Bessey and the founding
in 1895 of the first coherent group of American plant
ecologists, the Grassland School. Describes its basic
ideas (and the theory of succession in particular) and
contrasts it with the Chicago School. Identifies the
important role of the Agricultural Research Stations and
relates how the prolonged drought of 1933-1941 destroyed
the Clementsian theory and brought about the eventual
demise of the group.

129.   Tobey, Ronald C. "Theoretical Science and
       Technology in American Ecology." Technology and
       Culture 17(1976): 718-728.

Proposes the change from a static European
ecology to a dynamic American ecology came through the
efforts of Charles E. Bessey and his students, Roscoe
Pound and Frederic Clements, to control grassland
vegetation.

EMBRYOLOGY

130.   Adelmann, Howard B. Marcello Malpighi and the
       Evolution of Embryology. 5 vols. Ithaca:
       Cornell University Press, 1966.

Contains a detailed biography of Malpighi and an
extensive description of the development of embryology.
Includes valuable facsimiles and translations of
original sources, both published and manuscript, along
with commentary.

131.   Baxter, Alice Levine. "E.B. Wilson's
       'Destruction' of the Germ-Layer Theory." Isis
       68(1977): 363-374.

Describes E.B. Wilson's 1887-1890 research on the
embryology of annelids from which began his
dissatisfaction with the germ-layer theory after he
determined that there is no proof in the early cleavage
stage that homologies can be based on embryonic origins.
Documents that this dissatisfaction was one of the

reasons Wilson shifted his research to experimental embryology.

132. Benson, Keith R. "Problems of Individual Development: Descriptive Embryological Morphology in America at the Turn of the Century." Journal of the History of Biology 14(1981): 115-128.

Discusses the descriptive embryological work in morphology of William Keith Brooks (1848-1908) and his influence at Johns Hopkins in the 1880s and 1890s when morphology predominated over physiology. Suggests that morphology was the basis for experimental studies and that it uncovered problems and questions that physiology attempts to answer.

133. Blacher, Leonidas I. History of Embryology in Russia From the Middle of the Eighteenth to the Middle of the Nineteenth Century. Translated by Hosni Ibrahim Youssef, Boulos Abdel Malek, and Jane Maienschein. Washington, D.C.: Smithsonian Institution, 1982.

Surveys in detail the development of embryology in Russia from the Soviet point of view, focusing on Kaspar Friedrich Wolff and Karl Ernst von Baer, but including lesser known figures as well.

134. Bodemer, Charles W. "Regeneration and the Decline of Preformationism in Eighteenth Century Embryology." Bulletin of the History of Medicine 38(1964): 20-31.

Discusses the challenge that the confirmation of regeneration presented to preformationism.

135. Churchill, Frederick B. "Chabry, Roux, and the Experimental Method in Nineteenth Century Embryology." In Foundations of Scientific Method: The Nineteenth Century, edited by Ronald N. Giere and Richard S. Westfall, 161-205. Bloomington: Indiana University Press, 1973.

Contrasts two embryological experiments (lancing early blastomeres) by Laurent Marie Chabry and Wilhelm Roux from the viewpoint of their experimental techniques, their reasons for using experimentation, their employment of mechanical models, and their interpretation of their results. Considers also the research traditions from which they came.

136.   Churchill, Frederick B. "From Machine-Theory to Entelechy: Two Studies in Developmental Teleology." Journal of the History of Biology 2(1969): 165-185.

Discusses in detail the embryological research of Hans Driesch in the 1890s and of his transition from the mechanist view of development of his Analytische Theorie to the inclusion of a vital element (which he called "entelechy") in his "Die Lokalisation ..."

137.   Churchill, Frederick B. "The History of Embryology as Intellectual History." Journal of the History of Biology 3(1970): 155-181.

Compares and contrasts in an essay review five histories of embryology published in the 1960s, highlighting their intellectual history aspects.

138.   Fischer, Jean-Louis and Julian Smith. "French Embryology and the 'Mechanics of Development' from 1887 to 1910: L. Chabry, Y. Delage and E. Bataillon." History and Philosophy of the Life Sciences 6(1984): 25-39.

Describes briefly the embryological research of Laurent Chabry (1855-1893), Yves Delage (1854-1920), and Eugene Bataillon (1864-1953).

139.   Gould, Stephen Jay. Ontogeny and Phylogeny. Cambridge: Harvard University Press, 1977.

Traces and analyzes the history of the concepts of ontogeny and phylogeny from Aristotle to the present in a comprehensive manner.

140. Haraway, Donna Jeanne. Crystals, Fabrics and
      Fields: Metaphors of Organicism in Twentieth-
      Century Developmental Biology. New Haven: Yale
      University Press, 1976.

      Analyzes the contributions to experimental
embryology and cell biology of Ross G. Harrison, Joseph
Needham, and Paul Weiss in terms of their organicist
biology and a Kuhnian organicist paradigm.

141. Maienschein, Jane. "Experimental Biology in
      Transition: Harrison's Embryology, 1895-1910."
      Studies in History of Biology 6(1983): 107-127.

      Demonstrates that R.C. Harrison did not
experiment to emulate the physical sciences nor to
reject the morphological and descriptive approaches, but
rather developed slowly in response to internal and
external factors from a traditional descriptive approach
to a full experimental program. Emphasizes that his
German medical school training and his eschewal of
evolution as a stimulus or control on biological
research set him apart from his colleagues and
influenced his development. Cites different levels of
experimentalism and, using Harrison as an exemplar,
suggests that biologists' transition from descriptive to
experimental methods was gradual.

142. Maienschein, Jane. "Shifting Assumptions in
      American Biology: Embryology, 1890-1910."
      Journal of the History of Biology 14(1981): 89-
      113.

      Proposes a continua of methodology including
passive observation, systematic observation and
description, comparative description, use of
manipulative techniques or methods, and a fully
experimental approach. Describes the research of the
"Johns Hopkins Gang of Four," Edmund Beecher Wilson
(cytologist), Edwin Grant Conklin (evolutionary
biologist), Thomas Hunt Morgan (geneticist), and Ross
Granville Harrison (experimental embryologist) and
demonstrates that their research careers gradually moved
from observation to manipulative experimental
techniques, never fully abandoning any method.

143.   Manning, Kenneth R. Black Apollo of Science: The
       Life of Ernest Everett Just. New York: Oxford
       University Press, 1983.

       Discusses the scientific work of Everett Just and
the difficulties encountered in the scientific community
because of his race.

144.   Meyer, Arthur William. The Rise of Embryology.
       Stanford: Stanford University Press, 1939.

       Describes the history of the basic ideas in
embryology, including spontaneous generation,
epigenesis, preformation up through morphogenesis to
experimental embryology. Considered a standard early
source.

145.   Needham, Joseph. A History of Embryology. 2d.
       ed. revised with the assistance of Arthur
       Hughes. Cambridge: Cambridge University Press,
       1959.

       Details thoroughly the history of embryology from
antiquity through the eighteenth century. Remains a
basic, standard source.

146.   Oppenheimer, Jane M. "An Embryological Enigma in
       the Origin of Species." In Forerunners of
       Darwin: 1745-1859, edited by Bentley Glass, 292-
       322. Baltimore: Johns Hopkins Press, 1959.
       Reprinted 1968.

       Examines the influence of Karl Ernst von Baer's
embryological work on Darwin.

147.   Oppenheimer, Jane M. "Ernst Heinrich Haeckel as
       an Intermediary in the Transmutation of an
       Idea." Proceedings of the American
       Philosophical Society 126(1982): 347-355.

       Documents the transmission of the idea of
internal and external forces on the embryo from Wilhelm
Roux to Ernst Heinrich Haeckel to Hans Spemann. Traces
its early roots in the work of Kant, then Goethe, to
Roux.

148. Oppenheimer, Jane M. Essays in the History of
     Embryology and Biology. Cambridge: M.I.T.
     Press, 1967.

     Brings together thirteen essays, previously
presented or published elsewhere from 1940 to 1965,
covering major aspects of modern embryology.

149. Oppenheimer, Jane M. "Ross Harrison's
     Contributions to Experimental Embryology."
     Bulletin of the History of Medicine 40(1966):
     525-543.

     Demonstrates through a discussion of his research
that Harrison's real contribution to experimental
embryology and biology was the formulation of questions
to successfully approach the embryological problems of
his time, followed by his ability to devise simple
techniques to obtain answers to those questions.

150. Ospovat, Dov. "The Influence of Karl Ernst von
     Baer's Embryology, 1828-1859: A Reappraisal in
     the Light of Richard Owens' and William B.
     Carpenter's 'Paleontological Application of von
     Baer's Law.'" Journal of the History of Biology
     9(1976): 1-28.

     Details Karl Ernst von Baer's theory of
development and compares and contrasts it with the
theory of recapitulation. Provides examples of the
influence of von Baer's theory on his contemporaries.

151. Roe, Shirley A. "The Development of Albrecht von
     Haller's Views on Embryology." Journal of the
     History of Biology 8: 167-190.

     Follows the evolution of Albrecht von Haller's
ideas on embryological development from his student
spermaticist preformationism through epigenesis to ovist
preformationism.

152. Roe, Shirley A. Matter, Life, and Generation.
     Eighteenth-Century Embryology and the Haller-
     Wolff Debate. Cambridge: Cambridge University
     Press, 1981.

Examines the controversy between Albrecht von Haller, a preformationist, and Caspar Friedrich Wolff, an epigenesist, over embryological development. Covers the debate on the philosophical level, on the embryological level where the main issues were on the formation of blood vessels in the area vasculosa, the development of the heart, and on Haller's membrane-continuity proof of preformation, and covers Haller and Wolff's differences regarding the nature of scientific explanation. Includes translations of Wolff's letters to Haller.

153.   Roe, Shirley A. "Rationalism and Embryology: Caspar Friedrich Wolff's Theory of Epigenesis." Journal of the History of Biology 12(1979): 1-43.

Discusses the role Caspar Friedrich Wolff's philosophical beliefs, especially his rationalist beliefs, played in the formation of his embryological theories, which were in contrast to the largely preformationist theories of the period.

154.   Russell, E.S. Form and Function. A Contribution to the History of Animal Morphology. London: John Murray, 1916.

Remains a classic.

155.   Stephens, Trent D. "The Wolffian Ridge: History of a Misconception." Isis 73(1982): 254-259.

Documents the history of Charles Minot's incorrectly naming a part of the developing kidney "Wolffian Ridge" in 1892 when in fact "Wolffian Ridge" had already been used in 1868 by Wilhelm His in describing limb development. Records the resulting confusion and negative impact on modern embryologists.

156.   Willier, B.H., and Jane M. Oppenheimer, eds. Foundations of Experimental Embryology. 2d ed. New York: Hafner, 1974.

Translations and/or reprints of eleven classic embryological works by W. Roux, Hans Driesch, Edmund B.

Wilson, Theodor Boveri, Ross G. Harrison, Otto Warburg, Frank R. Lillie, C.M. Child, Hans Spemann and Hilde Mangold, and Johannes Holftreter. Each is accompanied by concise editor's comments.

ENDOCRINOLOGY

157. Beach, Frank A. "Historical Origins of Modern Research on Hormones and Behavior." Hormones and Behavior 15(1981): 325-376.

Surveys the major research contributions of zoologists, psychologists, neuroendocrinologists, physiologists, biologists, and neurophysiologists to the development of behavioral endocrinology. Covers 1849 to 1900, the "predisciplinary era," and 1900 to 1950s, the "formative era."

158. Brooks, Chandler McC. Humors, Hormones, and Neurosecretions: the Origins and Development of Man's Present Knowledge of the Humoral Control of Body Functions. Albany: State University of New York, 1962.

Contains eight chapters written with varying levels of historical content by four experts in the fields.

159. Medvei, Victor Cornelius. A History of Endocrinology. Lancaster, Boston: MTP Press, 1982.

Covers endocrinology comprehensively from prehistoric times to mid-twentieth century. Basic physiological research is reviewed, but the work is mainly a medical approach by a noted participant.

160. Raacke, I.D. "'The Die is Cast - I Am Going Home': The Appointment of Herbert McLean Evans as Head of Anatomy at Berkeley." Journal of the History of Biology 9(1976): 301-322.

Recounts Herbert McLean Evans' early research at Johns Hopkins, his decision to become Head of Anatomy at

Berkeley, his efforts to secure budget and staff to
build his department, and the genesis of his research
program there.

ENTOMOLOGY

161. Lindroth, Carl H. "Systematics Specializes
Between Fabricius and Darwin: 1800-1859." In
History of Entomology, edited by Ray F. Smith,
Thomas E. Mittler, and Carroll N. Smith, 119-
154. Palo Alto: Annual Reviews, 1973.

Identifies entomologists and their contributions
to systemization after specialization developed;
arranged chronologically under orders of insects.

162. Richard, G. "The Historical Development of
Nineteenth and Twentieth Century Studies on the
Behavior of Insects." In History of Entomology,
edited by Ray F. Smith, Thomas E. Mittler, and
Carroll N. Smith, 477-502. Palo Alto: Annual
Reviews, 1973.

Identifies major entomologists and their
contributions. Includes chart of the introduction and
significance of areas of behavioral research on insects.

163. Tuxen, S.L. "Entomology Systematizes and
Describes: 1700-1815." In History of
Entomology, edited by Ray F. Smith, Thomas E.
Mittler, and Carroll N. Smith, 95-118. Palo
Alto: Annual Reviews, 1973.

Surveys the contributions of eighteenth century
entomologists to insect identification, description, and
systematizing.

164. Wigglesworth, V.B. "The History of Insect
Physiology." In History of Entomology, edited
by Ray F. Smith, Thomas E. Mittler, Carroll N.
Smith, 203-228. Palo Alto: Annual Reviews,
1973.

Outlines contributions to insect physiology up to
1930.

ETHOLOGY

165. Beer, Colin Gordon. "Darwin, Instinct, and
Ethology." Journal of the History of the
Behavioral Sciences 19(1983): 68-80.

Cites the several different meanings of the term
instinct as it was used by Darwin.

166. Durant, J.R. "Innate Character in Animals and
Man: A Perspective on the Origins of Ethology."
In Biology, Medicine and Society, 1840-1940,
edited by Charles Webster, 157-192. Cambridge:
Cambridge University Press, 1981.

Discusses the development of ethology as one of
the precursors of sociobiology. Emphasizes the
influence of Oskar Heinroth in the 1920s and 1930s in
Germany. Covers also the roles of William Morton
Wheeler in America and Edmund Selous and Julian Huxley
in England.

167. Evans, Richard I. Konrad Lorenz, the Man and His
Ideas. New York: Harcourt Brace Jovanovich,
1975.

Contains an edited dialogue between the author
and Konrad Lorenz that is based on films and tapes
designed to introduce students to Lorenz's major
contributions. Includes reprints of four of Lorenz's
works, a bibliography of his works, an introduction by
the author, and a chapter by Donald Campbell with a
reply by Lorenz.

168. Gray, Philip Howard. "The Descriptive Study of
Imprinting in Birds from 1873-1953." Journal of
General Psychology 68(1963): 333-337.

Outlines briefly the earliest research on
imprinting, that done by Douglas Alexander Spalding, and
mentions some later field and laboratory studies also.

169.   Gray, Philip Howard.   "Douglas Alexander Spalding:
       The First Experimental Behaviourist."   Journal
       of General Psychology 67(1962): 299-307.

       Comments briefly on the unfavorable reception of
the research of Douglas Alexander Spalding, who
introduced the experimental study of learned and
instinctive animal behavior.

170.   Gray, Philip Howard.   "Early Animal Behaviorists:
       Prolegomenon to Ethology."   Isis 59(1968): 372-
       383.

       Describes the work of some amateur naturalists,
such as John James Audubon and Lewis Henry Morgan, who
contributed to the birth of the scientific study of
animal behavior.

171.   Kalikow, Theodora J.   "Konrad Lorenz's Ethological
       Theory: Explanation and Ideology, 1938-1943."
       Journal of the History of Biology 16(1983): 39-
       73.

       Sets some of Lorenz's ethological theories in the
scientific and political milieu of World War II Germany.
Focuses on his discussions of the degeneration of
society.

172.   Klopfer, Peter H., and Jack P. Hailman, eds.
       Function and Evolution of Behavior: An
       Historical Sample from the Pens of Ethologists.
       Reading, Mass.: Addison-Wesley, 1972.

       Contains twenty research papers significant in
the development of ethology arranged in groups under
four major issues of behavior.

173.   Klopfer, Peter H., and Jack P. Hailman, eds.
       Control and Development of Behavior: An
       Historical Sample from the Pens of Ethologists.
       Reading, Mass.: Addison-Wesley, 1972.

       Contains reprints of seventeen major ethological
papers, accompanied by brief introductions and
summaries.

174. Klopfer, Peter H. An Introduction to Animal
      Behavior: Ethology's First Century. 2d ed.
      Englewood Cliffs: Prentice-Hall, 1974.

         Surveys the development of ethology from 1850 to
      1970. Provides a chronological survey of early works.
      Covers the main thrust of the field, instinct, and all
      the concomitant areas: learning, physiological
      organization of behavior, hormones, sensation and
      perception, social life, and orientation. Includes an
      animal index and contains bibliographies of journals
      pertinent to the field and of basic articles on the
      various aspects of the field.

175. Nisbett, Alec. Konrad Lorenz. London: J.M. Dent,
      1976.

         Covers Lorenz's life and studies of animal
      behavior, particularly of jackdaws and gray leg geese,
      in a popular style. Based on letters and interviews
      with Lorenz and his associates; includes photographs.

176. Richards, Robert J. "The Emergence of
      Evolutionary Biology of Behaviour in the Early
      Nineteenth Century." British Journal for the
      History of Science 15(1982): 241-280.

         Explains eighteenth and nineteenth century
      theories of instinct and intelligence and analyzes the
      evolution of behavior and its role in species
      modification.

177. Richards, Robert J. "The Innate and the Learned:
      The Evolution of Konrad Lorenz's Theory of
      Instinct." Philosophy of Social Science
      4(1974): 111-133.

         Describes Konrad Lorenz's theory and research on
      innate behavior. Discusses how the theory has developed
      since its inception in 1935, always keeping in mind
      Lorenz's theoretical aims: taxonomy, explanation, and
      evolutionary understanding.

178. Richards, Robert J. "Instinct and Intelligence in
      British Natural Theology: Some Contributions to

Darwin's Theory of the Evolution of Behavior."
Journal of the History of Biology 14(1981): 193-230.

Recounts Darwin's changing ideas about instinct,
how they were influenced by the varying positions of
several natural theologians, and the role these changing
ideas played in the development of his theory of
evolution by natural selection.

179.   Sparks, John.   The Discovery of Animal Behaviour.
Boston: Little, Brown, 1982.

Surveys selectively in an engaging popular style
the development of the understanding of animal behavior
from early times to the present, with emphasis on the
twentieth century. Based on a television series and
profusely illustrated.

180.   Thorpe, W.H.   Learning and Instinct in Animals.
2d ed. London: Methuen, 1963.

Explains the basis of research in animal learning
and surveys observational and experimental studies in
animals from the protozoa to mammals.

181.   Thorpe, W.H.   The Origins and Rise of Ethology.
The Science of the Natural Behaviour of Animals.
London: Heinemann, 1979.

Surveys the origins of ethology from its roots in
natural history, its development in England, America,
and continental Europe during the late nineteenth and
early twentieth centuries through the post-war
establishment of research groups and laboratories. The
author participated in the twentieth century growth and
is strongest providing narrative and opinions of those
events. Contains good definitions of terms and model.

182.   Tinbergen, Niko.   The Animal in Its World:
Explorations of an Ethologist, 1932-1972. 2
vols. London: Allen & Unwin, University Press,
1972-1973.

Collects selectively Tinbergen's papers and translations of papers from 1932 through 1971 in two volumes covering "Field Studies" and "Laboratory Experiments and General Papers." Includes author's brief introductory notes commenting on his belief in the importance of studying behavior and its relevance to biological success.

183. Tinbergen, Niko. "Ethology." In Scientific Thought, 1900-1960, edited by Rom Harre, 238-268. Oxford: Clarendon Press, 1969.

Sketches the history of the development of ethology and its relation to other disciplines.

EUGENICS

184. Allen, Garland E. "Biology and Culture: Science and Society in the Eugenic Thought of H.J. Muller." BioScience 20(1970): 346-353.

Describes the American eugenics movement, Hermann Joseph Muller's lifelong interest in and suggested programs for eugenics. Discusses how his genetics research and his eugenics ideas interacted.

185. Roll-Hansen, Nils. "Eugenics before World War II: The Case of Norway." History and Philosophy of the Life Sciences 2(1980): 269-298.

Recounts the 1915 and after criticisms by biologists and medical men, led by Otto Louis Mohr, of the science and objectivity of the Norwegian eugenicists, led by Jan Alfred Mjoen. Covers the move to a moderate eugenics program by 1930 and the subsequent trend away from eugenics brought about by political experiences.

EVOLUTION

186. Adams, Mark B. "Towards a Synthesis: Population Concepts in Russian Evolutionary Thought, 1925-1935." Journal of the History of Biology

3(1970): 107-129.

Describes the attempts of Russian evolutionists
to effect a synthesis of genetic and naturalist
positions. Treats the research efforts of Sergei
Chetverikov, D.D. Romashov, and N.P. Dubinin to
identify, using wild populations, a genetic explanation
for the evolutionary effects of isolation and population
size.

187. Allen, Garland E. "Naturalists and
      Experimentalists: The Genotype and the
      Phenotype." Studies in the History of Biology
      3(1979): 179-209.

Discusses the different perspectives on questions
asked and answers accepted held by descriptive
naturalists and analytical experimentalists during the
nineteenth and early twentieth centuries. Cites the
Darwinians and Mendelians as examples, and discusses the
influence of Wilhelm Johannsen's 1909 distinction
between the genotype and the phenotype in reconciling
the two positions. Contains tables of background and
training of the two groups of biologists.

188. Allen, Garland E. "Thomas Hunt Morgan and the
      Problem of Natural Selection." Journal of the
      History of Biology 1(1968): 113-139.

Recounts anti-Darwinian natural selection
arguments related to the lack of a workable theory of
heredity through a case study of the development of
Thomas Hunt Morgan's concepts.

189. Appleman, Philip, ed. Darwin, a Norton Critical
      Edition. New York: Norton, 1970.

Arranges excerpts from primary sources and
commentary in six sections, including "Scientific
Opinion in Early Nineteenth Century" and "Darwin's
Influence on Science," with a final epilogue by the
editor.

190. Barlow, Nora, ed. Darwin and Henslow: the Growth
      of an Idea. Letters 1831-1860. London:

Bentham-Moxon Trust, John Murray, 1967.

Reproduces, with an introduction and notes, the correspondence between Charles Darwin and his Cambridge professor John Stevens Henslow from July 11, 1831, just before Darwin is invited to sail on the Beagle, to Dec. 10 [1860].

191. Barrett, Paul H., ed. The Collected Papers of
      Charles Darwin. 2 vols. Chicago: University of
      Chicago Press, 1977.

Contains reprints of 152 of Darwin's publications that have not previously appeared in book form and one previously unpublished paper. Includes a bibliography of those book publications and a bibliography of published descriptions of species that Darwin collected.

192. Beatty, John. "What's in a Word? Coming to Terms
      in the Darwinian Revolution." Journal of the
      History of Biology 15(1982): 215-239.

Analyzes the epistemological definition of "species" from nonevolutionary theories through the Darwinian Revolution of the 1920s and 1930s.

193. Beddall, Barbara G. "'Notes For Mr. Darwin':
      Letters to Charles Darwin from Edward Blyth at
      Calcutta." Journal of the History of Biology
      6(1973): 69-95.

Reviews Edward Blyth's contributions through his publications, his correspondence with Charles Darwin. Discusses the differences between his thinking on species and that of Darwin and Alfred Russel Wallace and the relative importance of his work to the development of the theory of evolution.

194. Beddall, Barbara G. "Wallace, Darwin, and Edward
      Blyth: Further notes on the Development of
      Evolutionary Theory." Journal of the History of
      Biology 5(1972): 153-158.

Includes a transcription of Edward Blyth's letter of December 8, 1865 to Charles Darwin in which he comments on Alfred Russel Wallace's paper "On the Law Which Has Regulated the Introduction of New-Species."

195. Beddall, Barbara G. "Wallace, Darwin, and the Theory of Natural Selection: A Study in the Development of Ideas and Attitudes." Journal of the History of Biology 1(1968): 261-323.

Discusses the development of Alfred Russel Wallace's ideas and his work. Recounts the events of 1858; re-evaluates and re-interprets contemporary sources.

196. Bell, P.R., ed. Darwin's Biological Work: Some Aspects Reconsidered. Cambridge: Cambridge University Press, 1959.

Discusses Darwin's work and its later development in five chapters ("Movement of Plants in Response to Light," "Paleontology and Evolution," "Natural Selection," "Animal Communication," "Cross- and Self-fertilization in Plants," and "Buffon, Lamarck, and Darwin: the Originality of Darwin's Theory of Evolution") written by five biologists.

197. Bell, Srilekha. "George Henry Lewes: A Man of His Time." Journal of the History of Biology 14(1981): 277-298.

Cites George Henry Lewes, who began his scientific career under the influence of German natural philosophy and then became a Darwinian, as an exemplar of change in scientific ideas in the nineteenth century.

198. Bendall, D.S., ed. Evolution: From Molecules to Men. Cambridge: Cambridge University Press, 1983.

Includes four chapters on evolutionary history: "Darwin, Intellectual Revolutionary" by Ernst Mayr, "The Development of Darwin's General Biological Theorizing" by M.J.S. Hodge, "Darwin and the Nature of Science" by D.L. Hull, and "The Several Faces of Darwin: Materialism

in Nineteenth and Twentieth Century Evolutionary Theory"
by G.E. Allen.

199. Bowler, Peter J. "The Changing Meaning of
      'Evolution.'" Journal of the History of Ideas
      36(1975): 95-114.

      Explains the differing meanings of the word
evolution in eighteenth century embryology, in
transmutation, as used by Herbert Spencer, and modern
usage after Darwin.

200. Bowler, Peter J. "Darwinism and the Argument from
      Design: Suggestions for a Reevaluation."
      Journal of the History of Biology 10(1977): 29-
      43.

      Discusses the idealist argument from design in
England at the time of Darwin.

201. Bowler, Peter J. Eclipse of Darwinism.
      Baltimore: Johns Hopkins University Press, 1983.

      Analyzes the reasons for the resurgence of
opposition to natural selection as a mechanism for
evolution around 1900. Outlines the conceptual and
historical issues, the religious and philosophical
implications, and then discusses the alternative
theories thoroughly.

202. Browne, Janet. "Darwin's Botanical Arithmetic and
      the 'Principle of Divergence,' 1854-1858."
      Journal of the History of Biology 13(1980): 53-
      89.

      Discusses Darwin's use of botanical arithmetic in
Charles Darwin's studies of the relationships of genera,
species, and varieties and explains the important role
his barnacle studies and use of botanical arithmetic
played in his discovery of the "principle of
divergence."

203. Burkhardt, Richard W., Jr. "The Inspiration of
      Lamarck's Belief in Evolution." Journal of the

History of Biology 5(1972): 413-438.

Outlines Jean Baptiste Lamarck's overall
biological thought which he developed early. Proposes
that his ideas on evolution arose from those beliefs,
from his work as a conchologist and the differences
between fossil and contemporary shells, and from his
ideas on the origin of life and his work with the monad.

204.   Burkhardt, Richard W., Jr. "Lamarck, Evolution
        and the Politics of Science." Journal of the
        History of Biology 3(1970): 275-298.

Analyzes the reasons for the poor reception of
Jean Baptiste Lamarck's evolutionary theory. Suggests
that Lamarck's personal view of science and the
scientific community was a factor, along with those
factors traditionally cited - hostile scientific views
or the insufficiency of Lamarck's arguments.

205.   Burkhardt, Richard W., Jr. The Spirit of System:
        Lamarck and Evolutionary Biology. Cambridge:
        Harvard University Press, 1977.

Discusses the development of Lamarck's research,
his contributions to botany and invertebrate zoology,
his theories, particularly his evolutionary theory, and
their reception in the scientific context of the late
eighteenth and early nineteenth centuries.

206.   Bynum, W.F. "Charles Lyell's Antiquity of Man and
        Its Critics." Journal of the History of Biology
        17(1984): 153-188.

Describes the contemporary reception of Charles
Lyell's Antiquity of Man, particularly the controversies
with Richard Owen, Hugh Falconer, John Lubbock, and
others that arose. Re-evaluates its role in the
establishment of the discipline of anthropology.

207.   Conry, Yvette. De Darwin au Darwinisme: Science
        et Idéologie. Paris: Vrin, 1983.

Contains sixteen diverse papers presented at the Congrès International pour le Centenaire de la Mort de Darwin held in Paris, in September, 1982.

208.   Conry, Yvette.   L'Introduction du Darwinisme en France en XIXe Siècle.   Paris: J. Vrin, 1974.

Discusses the reception of Darwinism in France and includes a useful bibliography of published commentary from the publication of the (mis)translation of Darwin's The Origin of Species to the turn of the century.

209.   Cornell, John F.   "From Creation to Evolution: Sir William Dawson and The Idea of Design in the Nineteenth Century."   Journal of the History of Biology 16(1983): 137-170.

Discusses the nineteenth century dichotomy between design by special creation and evolution by natural selection and the effects of attempts to clarify these issues on the concepts of the paleontologist William Dawson.

210.   deBeer, Sir Gavin.   Charles Darwin: Evolution by Natural Selection.   London: Nelson, 1963.

Describes in popular style the development of Charles Darwin's concepts and the reception of the Origin of the Species.

211.   Desmond, Adrian J.   "Robert E. Grant: The Social Predicament of a Pre-Darwinian Transmutationist."   Journal of the History of Biology 17(1984): 189-223.

Identifies several reasons, including his Lamarckism, for the collapse of Robert Edmond Grant's promising scientific career.

212.   Di Gregorio, Mario A.   "The Dinosaur Connection: A Reinterpretation of T.H. Huxley's Evolutionary View."   Journal of the History of Biology 15(1982): 397-418.

Assigns the major influences on Thomas Henry
Huxley's work to the German scientific community,
particularly to Karl Ernst von Baer's embryological
typology in the early 1850s and to Ernst Haeckel's
Generelle Morphologie der Organismen of 1866.

213. Di Gregorio, Mario A. "Order or Process of
      Nature: Huxley's and Darwin's Different
      Approaches to Natural Sciences." History and
      Philosophy of the Life Sciences 3(1981): 217-
      241.

Explains the reasons for Thomas Henry Huxley's
support of Charles Darwin's theory and the reasons why
Huxley did not agree with it in some areas.

214. Egerton, Frank N. "Humboldt, Darwin, and
      Population." Journal of the History of Biology
      3(1970): 325-360.

Documents and evaluates Humboldt's influences on
Darwin's ideas about human and animal populations.
Notes the importance of Darwin's biological rather than
economic orientation when studying animal populations.

215. Eiseley, Loren. "Charles Darwin, Edward Blyth,
      and the Theory of Natural Selection."
      Proceedings of the American Philosophical
      Society 103(1959): 94-158.

Suggests that Edward Blyth played a more
important role in the development of the theory of
evolution than that attributed to him. Includes
reprints of three of Blyth's papers.

216. Eiseley, Loren. Darwin's Century: Evolution and
      the Men Who Discovered It. New York: Doubleday,
      1958.

Surveys concepts of "evolution" of nineteenth
century natural historians pre and post Darwin.

217. Evans, L.T. "Darwin's Use of the Analogy between
      Artificial and Natural Selection." Journal of

the History of Biology 17(1984): 113-140.

Demonstrates the importance of Charles Darwin's analogy between artificial and natural selection to his theory of evolution. Suggests Darwin's use of the analogy prepared him for his insight while reading Malthus, for collecting supporting evidence, and for presenting his theory.

218. Farber, Paul Lawrence. "Buffon and the Concept of Species." Journal of the History of Biology 5(1972): 259-284.

Explains the development of Buffon's concept of species. Covers its relationship to and impact on eighteenth century biology and general thought.

219. Farley, John. "The Initial Reactions of French Biologists to Darwin"s Origin of the Species." Journal of the History of Biology 7(1974): 275-300.

Explains why the concepts in Charles Darwin's Origin of the Species were rejected in France and briefly contrasts this with their German reception.

220. Freeman, R.B. The Works of Charles Darwin: An Annotated Bibliographical Handlist. 2d ed. Folkestone, Kent: Dawson, 1977.

Lists, with brief annotations, all Darwin's works. Contains an introductory essay on bibliographical issues.

221. George, Wilma. Biologist Philosopher: A Study of the Life and Writing of Alfred Russel Wallace. London: Abelard-Schuman, 1964.

Discusses Alfred Russel Wallace's contributions to biology, including evolution and zoogeography.

222. Ghiselin, Michael T. The Triumph of the Darwinian Method. Berkeley: University of California Press, 1969.

Analyzes Charles Darwin's concepts and method of research throughout all of his major work. Emphasizes the importance of his methodology to his success.

223. Ghiselin, Michael T. "Two Darwins: History Versus Criticism."˙ Journal of the History of Biology 9(1976): 121-132.

Compares and contrasts the work of Charles Darwin and Erasmus Darwin.

224. Glass, Bentley. "Evolution and Heredity in the Nineteenth Century." In Medicine, Science and Culture: Historical Essays in Honor of Owsei Temkin, edited by Lloyd G. Stevenson and Robert P. Multhauf, 209-246. Baltimore: Johns Hopkins Press, 1968.

Considers "the great central themes or ideas of modern biology," including evolution and heredity.

225. Glass, Bentley, Owsei Temkin, William L. Straus, eds. Forerunners of Darwin: 1745-1859. Baltimore: Johns Hopkins Press, 1959. Reprinted 1968.

Contains fifteen essays discussing the eighteenth and nineteenth century ideas that provided the rich context in which Charles Darwin's Origin of the Species could be written.

226. Glick, Thomas F., ed. The Comparative Reception of Darwinism. Austin: University of Texas Press, 1972.

Contains papers presented at a conference on the comparative reception of Darwinism. Includes chapters on England, Scotland, Germany, France, United States, Russia, Netherlands, Spain, Mexico, and the Islamic world discussing the scientific, religious, economic, and social reception of Darwinism and bibliographic points of editions, translations, and reviews.

227. Gould, Stephen Jay. "Agassiz's Marginalia in
      Lyell's Principles, or the Perils of Uniformity
      and the Ambiguity of Heroes." Studies in
      History of Biology 3(1979): 119-138.

      Argues forcefully for a new interpretation of the
uniformitarian-catastrophist debate. Cites the marginal
annotations made by Louis Agassiz in his copy of Charles
Lyell's Principles as the stimulus for re-examining the
two positions and identifying their common ground.

228. Gould, Stephen Jay. "Dollo on Dollo's Law:
      Irreversibility and the Status of Evolutionary
      Laws." Journal of the History of Biology
      3(1970): 189-212.

      Explains Louis Dollo's law and various
misinterpretations of it. Includes a translation of
Dollo's 1893 "The Laws of Evolution" and a short
bibliography of Dollo's work on irreversibility.

229. Gould, Stephen Jay. "Trigonia and the Origin of
      Species." Journal of the History of Biology
      1(1968): 41-56.

      Recounts the story of the discovery in 1802 of a
"living fossil," Trigonia. Describes the responses of
Lamarck, James Parkinson, and Louis Agassiz, whose
existing theories of life had to incorporate the anomaly
of the Trigonia.

230. Greene, John C. The Death of Adam. Evolution and
      Its Impact on Western Thought. Ames: Iowa State
      University Press, 1959.

      Covers at an introductory level ideas about
nature and about evolution that existed between 1690 and
Charles Darwin's publication of his Origin of the
Species and Descent of Man.

231. Grene, Marjorie, ed. Dimensions of Darwinism:
      Themes and Counterthemes in Twentieth Century
      Evolutionary Theory. Cambridge: Cambridge
      University Press, 1983.

Collects twelve papers presented at a 1981 conference on the history of twentieth century evolutionary theory, the evolutionary synthesis in particular.

232. Grinnell, George. "The Rise and Fall of Darwin's First Theory of Transmutation." Journal of the History of Biology 7(1974): 259-273.

Suggests that Charles Darwin proposed various theories and then examined his data to see if it would fit. Describes the first such theory and the reasons for its collapse.

233. Harrison, James. "Erasmus Darwin's View of Evolution." Journal of the History of Ideas 32(1971): 247-264.

Extends the view of Erasmus Darwin's evolution as Lamarckian by providing from his writings examples of his evolutionary thought in its fullest.

234. Herbert, Sandra. "Darwin, Malthus, and Selection." Journal of the History of Biology 4(1971): 209-217.

Interprets new evidence (recently found pages from 'Notebooks on Transmutation of Species') on when Darwin read Malthus and on how much Darwin was influenced by Malthus.

235. Himmelfarb, Gertrude. Darwin and the Darwinian Revolution. Gloucester, Mass.: Peter Smith, c1962, 1967.

Describes the development of the theory of evolution by natural selection in its scientific context, analyzes the theory, and discusses its effect on science, religion, politics, and society.

236. Hodge, M.J.S. "Darwin and the Laws of the Animate Part of the Terrestrial System (1835-1837): On the Lyellian Origins of His Zoonomical Explanatory Program." Studies in History of

Biology 6(1983): 1-106.

Proposes an historiographic approach through a
reconstruction and analyses of Charles Darwin's
development from 1835 to 1837 of an explanatory program
while retaining the explanatory traditions of Charles
Lyell.

237. Hodge, M.J.S. "The Universal Gestation of Nature:
      Chambers' 'Vestiges and Explanations.'" Journal
      of the History of Biology 5(1972): 127-151.

Analyzes Robert Chambers' intentions and ideas on
organic diversity as they appeared in his Vestiges of
Natural History of Creation (especially his explanatory
preface to the 1853 edition) and Explanations: A Sequel
to the Vestiges. Uses Chambers' interpretation of the
Galapagos data to highlight the difference between his
developmental argument and the Darwinian's common
descent argument.

238. Hull, David L. Darwin and His Critics. The
      Reception of Darwin's Theory of Evolution by the
      Scientific Community. Cambridge: Harvard
      University Press, 1973.

Documents the scientific reception of Charles
Darwin's theory of evolution by reprinting (in
translation where necessary) sixteen reviews written in
Great Britain, the United States, France and Germany by
his scientific contemporaries. Accompanying each review
are quotations from the correspondence generated by it
and commentary on the author of the review and its
context.

239. Irvine, William. Apes, Angels and Victorians:
      Darwin, Huxley, and Evolution. Cleveland:
      Meridian, c1955, 1968.

Describes in a popularized style the development
of Charles Darwin's and Thomas Henry Huxley's work;
discusses the reception of the Origin and Huxley's
defense of it; follows both thru post-Origin careers.

240. Keith, Sir Arthur. Darwin Revalued. London:
      Watts, 1955.

      Describes in a popular style Charles Darwin's
life, the formulation of his theory of evolution, and
his writings, followed by chapters on such things as his
mother's influence and his family tree.

241. Kingsland, Sharon. "Abbott Thayer and the
      Protective Coloration Debate." Journal of the
      History of Biology 11(1978): 223-244.

      Reviews the variety of explanations offered to
account for animal coloration at the end of the
nineteenth century. Discusses Abbott Handerson Thayer's
ideas of coloration for concealment, his law of counter-
shading, and the controversy surrounding his ideas.

242. Kohn, David. "Theories to Work By: Rejected
      Theories, Reproduction, and Darwin's Path to
      Natural Selection." Studies in History of
      Biology 4(1980): 67-170.

      Reconstructs Darwin's thinking on the
transformist themes of adaptation, extinction, role of
reproduction, and geographic distribution. Analyzes
Darwin's explanations of the origin and extinction of
species from 1835 to 1838 through a series of five
theoretical frameworks within which Darwin worked.
Emphasizes Darwin's ideas on the role of reproduction in
the construction and rejection of the theories.

243. Lesch, John E. "The Role of Isolation in
      Evolution: George J. Romanes and John T.
      Gulick." Isis 66(1975): 483-503.

      Focuses on the contributions of George J. Romanes
and John T. Gulick in clarifying problems with Charles
Darwin's theory of natural selection and in stimulating
debate over possible solutions, particularly the role of
isolation.

244. Limoges, Camille. La Sélection Naturelle: Étude
      sur la Première Constitution d'un Concept (1837-

1859). Paris: Presses Universitaires de France,
1970.

Traces chronologically the development of Charles
Darwin's theory of evolution by natural selection from
1837 to 1859 using Darwin's manuscripts. Argues that
Darwin's artificial selection analogy does not play an
important role in his developing thought on natural
selection.

245. Lurie, Edward. "Louis Agassiz and the Idea of
      Evolution." Victorian Studies 3(1959-60): 87-
      108.

Identifies influences on the development of Louis
Agassiz's thought, including his studies at the
University of Munich and adherence to Cuvier. Analyzes
Agassiz's original reception and long continued vehement
opposition to Darwin's theory of evolution. Describes
Agassiz's eventual attempt and failure to understand
that theory.

246. McKinney, H. Lewis, ed. Lamarck to Darwin:
      Contributions to Evolutionary Biology 1809-1859.
      Lawrence, Kansas: Coronado, 1971.

Reprints ten classic papers (some in translation)
on variation, evolution, and natural selection from the
fifty years preceding the publication of Origin of the
Species. Includes works by J.B.P.A. Lamarck, William
Charles Wells, Patrick Matthew, Charles Lyell, Edward
Blyth, Robert Chambers, Alfred Russel Wallace and
Charles Darwin.

247. McKinney, H. Lewis. Wallace and Natural
      Selection. New Haven: Yale University Press,
      1972.

Describes the independent development of Alfred
Russel Wallace's ideas on evolution by natural
selection, focusing on the 1845 to 1858 period and
utilizing manuscript sources.

248. McKinney, H. Lewis. "Wallace's Earliest
      Observations on Evolution: 28 December 1845."

Isis 60(1969): 370-373.

Reproduces the clearly dated original letter of Alfred Russel Wallace to Henry Walter Bates which contains his earliest observations on evolution.

249. Mayr, Ernst. "Agassiz, Darwin and Evolution." Harvard Library Bulletin 13(1959): 165-194.

Reviews Louis Agassiz's theory and his opposition to the theory of evolution.

250. Mayr, Ernst, and William B. Provine, eds. The Evolutionary Synthesis. Perspectives on the Unification of Biology. Cambridge: Harvard University Press, 1980.

Contains papers presented by historians and scientists at a conference on the history of the evolutionary synthesis of 1920-1950. Papers are grouped by discipline (genetics, cytology, embryology, systematics, botany, paleontology, morphology) and by country (Soviet Union, Germany, France, England, United States) with a final section on interpretive issues and biographical essays.

251. Mayr, Ernst. The Growth of Biological Thought: Diversity, Evolution, and Inheritance. Cambridge, Mass.: The Belknap Press of Harvard University Press, 1982.

Analyzes the major issues in the historical development of evolutionary biology; a modern classic.

252. Mayr, Ernst. "Lamarck Revisited." Journal of the History of Biology 5(1972): 55-94.

Analyzes in detail Lamarck's evolutionary thought based on his Philosophie Zoologique from which numerous quotations are included.

253. Millhauser, Milton. Just before Darwin: Robert Chambers and "Vestiges." Middletown, Conn.: Wesleyan University Press, 1959.

Discusses Robert Chambers and the reception of his anonymously published book the <u>Vestiges of the Natural History of Creation</u>, an attempt to synthesize pre-Darwinian evolutionary ideas.

254.   Norton, Bernard J.   "The Biometric Defense of Darwinism."   <u>Journal of the History of Biology</u> 6(1973): 283-316.

Describes W.F.R. Weldon's research on death rates in nature (using crabs and snails) and their correlation with very small variations in order to refute William Bateson and support Charles Darwin's concepts of natural selection.   Includes discussion of Karl Pearson's disproof of Francis Galton's theory of perpetual regression.

255.   Olby, Robert C.   <u>Charles Darwin</u>.   London: Oxford University Press, 1967.

Describes very briefly, in an introductory way, Charles Darwin's education, voyage on the Beagle, the birth of his theory of evolution and the writing of <u>Origin of the Species</u>.

256.   Oldroyd, D.R.   <u>Darwinian Impacts.   An Introduction to the Darwinian Revolution</u>.   Atlantic Highlands, N.J.: Humanities Press, 1980.

Discusses the main natural history themes prior to Darwin, Darwin's life and work, with an analysis of his theory of evolution by natural selection, and post Darwin evolutionary theories.   Describes the impact of Darwin's work on politics, theology, philosophy, psychology, anthropology, literature, and music.

257.   Ospovat, Dov.   <u>The Development of Darwin's Theory: Natural History, Natural Theology and Natural Selection, 1838-1859</u>.   Cambridge: Cambridge University Press, 1981.

Covers the development of Darwin's theory of natural selection between 1838 and 1859.   Discusses Darwin's early positions and the influences contributing to significant changes in his theory that took place between 1844 and 1859, especially the elimination of some ideas based on natural theology and the development of such ideas as relative adaptation.

258.   Pancaldi, Guiliano. <u>Darwin in Italia: Impresa Scientifica e Frontiere Culturali</u>. Bologna: Il Mulino, 1983.

Discusses the natural sciences in Italy in five sections, particularly the debate over species.

259.   Rheinberger, Hans-Jorg, and Peter McLaughlin. "Darwin's Experimental Natural History." <u>Journal of the History of Biology</u> 17(1984): 345-368.

Proposes that Darwin saw "natural experiments" and artificial selection in domestic animals as experimental approaches for the theory of evolution. Recounts the critiques of the latter model by William Hopkins and Thomas Henry Huxley.

260.   Richardson, R. Alan. "Biogeography and the Genesis of Darwin's Ideas on Transmutation." <u>Journal of the History of Biology</u> 14(1981): 1-41.

Discusses Darwin's early work on biogeography based on Cambridge University's collection of his personal annotated books and articles, and his personal papers, which include correspondence, drafts, diaries, reading notes, etc. Suggests that natural selection was the model chosen by Darwin because of the large body of information cumulated from multiple sources during his studies of biogeography which provided so many examples of the relationship of species to their environment.

261.   Ridley, Mark. "Coadaptation and the Inadequacy of Natural Selection." <u>British Journal of the History of Science</u> 15(1982): 45-68.

Analyzes the nineteenth century views that natural selection could not account for coadaptation; explains alternative explanations of coadaptation posited by leading scientists.

262. Rogers, James Allen. "The Reception of Darwin's
      Origin of the Species by Russian Scientists."
      Isis 64(1973): 484-503.

      Identifies Russian scientists Christian H.
Pander, Alexander Keyserling, Karl Ernst von Baer, Karl
Rouillier, Vladimir and Alexander Kovalevsky, and
Kliment Arkadevich Timiriazev and describes their work
and their response to Origin of the Species.

263. Rudwick, Martin J.S. "Charles Darwin in London:
      The Integration of Public and Private Science."
      Isis 73(1982): 186-206.

      Maps the social and cognitive topography of
geology indicating Darwin's place on it from 1837 to
1842 while he was living in London after his Beagle
voyage. Maps Darwin's work during that same time on a
scale of relative privacy with his geological
participation from the first map representing the most
public of his work, his work on the origin of the
species private and his work on man and mind most
private.

264. Ruse, Michael. "Charles Darwin and Artificial
      Selection." Journal of the History of Ideas
      36(1975): 339-350.

      Argues that Charles Darwin's knowledge of
artificial selection and the analogy between it and
natural selection were definitely crucial elements in
his discovery of the theory of evolution based on
natural selection, because it was this prior knowledge
that he re-evaluated when he read Malthus.

265. Ruse, Michael. "Charles Darwin's Theory of
      Evolution: An Analysis." Journal of the History
      of Biology 8(1975): 219-241.

      Discusses the nature of the arguments of Charles
Darwin's theory of evolution.

266. Ruse, Michael. The Darwinian Revolution.
      Chicago: The University of Chicago Press, 1979.

60   Evolution

Surveys concepts of the origins of organisms from 1830 to 1875, a classic discussion of the scientific, philosophical, and religious aspects in context.

267. Sanford, William F., Jr. "Dana and Darwinism." Journal of the History of Ideas 26(1965): 531-546.

Describes James Dwight Dana's non-developmental position, his initial open-minded response to the Origin of the Species, and his eventual support of many Darwinian evolutionary concepts.

268. Schwartz, Joel S. "Charles Darwin's Debt to Malthus and Edward Blyth." Journal of the History of Biology 7(1974): 301-318.

Determines that Edward Blyth's work was not an important prelude to Darwin's theory of natural selection. Discusses the order in which Darwin read Blyth and Malthus and the level of Malthus' influence on Darwin.

269. Schwartz, Joel S. "Darwin, Wallace, and the Descent of Man." Journal of the History of Biology 17(1984): 271-290.

Analyzes the influences, including phrenology, spiritualism, and Owenite socialism, that effected Alfred Russel Wallace's change of view on natural selection in man.

270. Schweber, Silvan S. "The Origin of the Origin Revisited." Journal of the History of Biology 10(1977): 229-316.

Argues that Darwin had the basis for his theory by August, 1838, and analyzes in detail the influences of that period, including Darwin's reading of Brewster's review, Adam Smith, Quetelet, and Malthus, and changes in Darwin's religious beliefs.

271. Secord, J.A. "Nature's Fancy: Charles Darwin and
      the Breeding of Pigeons." Isis 72(1981): 163-
      186.

      Discusses Charles Darwin's immersion in pigeon
breeding and the fancying world during his study of
domestication and artificial selection.

272. Sheets-Pyenson, Susan. "Darwin's Data: His
      Reading of Natural History Journals, 1837-1842."
      Journal of the History of Biology 14(1981): 231-
      248.

      Describes Charles Darwin's reading of the
Magazine of Zoology and Botany, the Magazine of Natural
History, and the Annals of Natural History, his method
of annotating them, and their influence on his work.

273. Stauffer, Robert Clinton, ed. Charles Darwin's
      Natural Selection: Being the Second Part of His
      Big Species Book Written From 1856 to 1858.
      Cambridge: Cambridge University Press, 1975.

      Publishes, with expert editorial contributions,
Charles Darwin's natural selection manuscript.

274. Sulloway, Frank J. "Darwin and His Finches: the
      Evolution of a Legend." Journal of the History
      of Biology 15(1982): 1-53.

      Recounts the story of Charles Darwin's Galapagos
Island finches, the stimulus they were supposed to have
provided him, and their use as a textbook example.
Disproves the legend by a thorough analysis of Darwin's
manuscript (voyage diary, notebooks, etc.) and published
materials that establishes the retrospective nature of
Darwin's understanding.

275. Sulloway, Frank J. "Further Remarks on Darwin's
      Spelling Habits and the Dating of Beagle Voyage
      Manuscripts." Journal of the History of Biology
      16(1983): 361-390.

      Modifies his table of variations in Charles

Darwin's spelling between 1832 and 1836 on his voyage. Suggests additional historiographic uses.

276. Sulloway, Frank J. "Geographic Isolation in Darwin's Thinking: The Vicissitudes of a Crucial Idea." Studies in History of Biology 3(1979): 23-65.

Analyzes three stages of changes in Charles Darwin's thinking regarding geographical isolation and discusses the reasons behind them.

277. Temkin, Owsei. "The Idea of Descent in Post-Romantic German Biology: 1848-1858." In Forerunners of Darwin: 1745-1859, edited by Bentley Glass, 323-355. Baltimore: Johns Hopkins Press, 1959. Reprinted 1968.

Analyzes the lack of interest in evolution in Germany prior to the publication of Origin of the Species. Discusses the ideas of those scientists who were interested in evolution.

278. Vergata, Anotonello La, ed. L'Evoluzione Biologica: Da Linneo A Darwin. Torino: Loescher Editore, 1979.

Reprints classic texts from the major themes of eighteenth and nineteenth century evolutionary biology, such as the chain of being, the fixity of species, and natural selection.

279. Vorzimmer, Peter J. Charles Darwin: The Years of Controversy. The Origin of Species and Its Critics, 1859-1882. Philadelphia: Temple University Press, 1970.

Covers the twenty-three years after the publication of the Origin of the Species. Focuses on the concepts of the first five chapters and in particular on the mechanism of natural selection. Discusses criticisms by Darwin's contemporaries and his responses to them.

280. Vorzimmer, Peter J. "Darwin, Malthus, and the
     Theory of Natural Selection." Journal of the
     History of Ideas 30(1969): 527-542.

     Suggests that traditional interpretations of
Malthus' influence on Darwin are inadequate. Analyzes
Darwin's work from his return from the Beagle voyage to
his reading of Malthus two years later. Details
Darwin's "awareness" and "unawareness" of key points at
the time of his reading of Malthus and proposes an
alternative interpretation.

281. Vorzimmer, Peter J. "Darwin's Questions About the
     Breeding of Animals (1839)." Journal of the
     History of Biology 2(1969): 269-281.

     Discusses and reprints (only two copies known to
exist) an 1839 eight page privately printed work of
Charles Darwin which contains forty-eight questions
about the breeding of animals.

282. Vorzimmer, Peter J. "The Darwin Reading Notebooks
     (1838-1860)." Journal of the History of Biology
     10(1977): 107-153.

     Contains a chronological list of books, of both
scientific and leisure nature, that Darwin indicated he
had read.

283. Vorzimmer, Peter J. "An Early Darwin Manuscript:
     The 'Outline and Draft of 1839.'" Journal of
     the History of Biology 8(1975): 191-218.

     Analyzes a newly located Darwin manuscript that
contains his evolutionary argument and establishes an
approximate date in 1839. Includes a transcript of the
manuscript.

284. Wells, Kentwood D. "The Historical Context of
     Natural Selection: The Case of Patrick Matthew."
     Journal of the History of Biology 6(1973): 225-
     258.

     Reviews Patrick Matthew's ideas on natural
selection as they appear in his appendix to Naval Timber

and Arboriculture and in Emigration Fields, a previously
overlooked source. Distinguishes between Matthew's
ideas and Darwin's work.

285. Wilson, Leonard G., ed. Sir Charles Lyell's
      Scientific Journals on the Species Question.
      New Haven and London: Yale University Press,
      1970.

Contains reproduction of Charles Lyell's
scientific journals I-VII, 1855-1861, with explanatory
notes at the end of each. Includes a description of
Lyell's pre-1855 thought and of the species question.

286. Winsor, Mary Pickard. "Louis Agassiz and the
      Species Question." Studies in History of
      Biology 3(1979): 80-117.

Modifies early views of Agassiz's position on the
species question and the basis for it. Diminishes the
importance of religion and emphasizes the role that his
scientific experiences with species, especially fish and
turtles, had on his concepts of the identifiability,
fixity, and arrangements of species.

287. Young, Robert Maxwell. "Malthus and the
      Evolutionists: The Common Context of Biological
      and Social Theory." Past and Present 43(1969):
      109-145.

Documents the varying social and scientific
responses elicited by Thomas Malthus' Essay on the
Principle of Population; focuses on William Paley,
Thomas Chalmers, Charles Darwin, Alfred Russel Wallace,
Herbert Spencer, Karl Marx, and Engels.

GENERATION AND REPRODUCTION

288. Bowler, Peter J. "Bonnet and Buffon: Theories of
      Generation and the Problem of Species." Journal
      of the History of Biology 6(1973): 259-281.

Explains clearly and compares and contrasts the

concepts of generation held by Charles Bonnet and
Georges Louis Leclerc, comte de Buffon.

289.   Churchill, Frederick B. "Sex and the Single
        Organism: Biological Theories of Sexuality in
        Mid-Nineteenth Century." Studies in History of
        Biology 3(1979): 139-177.

        Illuminates the period 1830-1875 during which
attention turned to invertebrates, lower plants, and non
bi-sexual reproduction. Examines alternation of
generations and parthenogenesis; covers the research of
Johann Steenstrup, Richard Owen, Johannes Dzierzon, Karl
von Siebold, Rudolph Leuckart and others and the growth
of knowledge about methods of asexual reproduction.

290.   Cole, Francis Joseph. Early Theories of Sexual
        Generation. Oxford: Clarendon, 1930.

        Describes fully the history of concepts of the
spermatozoa, the preformation doctrine, epigenesis, and
early theories of fertilization and development.
Standard early source.

291.   Eddy, J.H. "Buffon, Organic Alterations, and
        Man." Studies in History of Biology 7(1984): 1-
        45.

        Describes the aspects of Buffon's philosophy of
science and the natural realm, his theory of generation
and the concept of species, and his theory of organic
alteration which together explain his ideas of organic
alteration within the human species.

292.   Farley, John. Gametes and Spores: Ideas About
        Sexual Reproduction, 1750-1914. Baltimore:
        Johns Hopkins University Press, 1982.

        Discusses thoroughly the concepts of and the
research on sexual reproduction in plants and animals
from mid-eighteenth century to 1914. Includes a chapter
on nineteenth century social views of sex and a
postscript on the teaching of outmoded information in
schools.

293. Farley, John. The Spontaneous Generation
      Controversy From Descartes to Oparin.
      Baltimore: Johns Hopkins University Press, 1977.

      Presents a thorough, well documented analysis of
the theories of and research on spontaneous generation
in the context of scientific thought. Covers the
seventeenth century to mid-twentieth century, with focus
on the nineteenth century.

294. Farley, John. "The Spontaneous Generation
      Controversy (1700-1860): The Origin of Parasitic
      Worms." Journal of the History of Biology
      5(1972): 95-125.

      Reviews theories of the spontaneous origin of
worms, including both the internalist and the
externalist positions. Explains the difficulty of using
experimental methods to decide the question. Discusses
the theories of the founders of the German School,
Marcus E. Bloch, Johann Goeze, Johann Bremser, and Karl
Rudolphi. Suggests that the reasons for rejecting
spontaneous generation included the rejection of
Naturphilosophie, the rise of pathological anatomy, and
Ehrenberg's infusorian research.

295. Farley, John. "The Spontaneous Generation
      Controversy (1859-1880): British and German
      Reactions to the Problem of Abiogenesis."
      Journal of the History of Biology 5(1972):
      285-319.

      Differentiates the mid-nineteenth century
controversy over the form of spontaneous generation,
abiogenesis, which was stimulated by the publication of
Darwin's Origin, from the earlier over heterogenesis.
Contrasts the British and German responses.

296. Gasking, Elizabeth B. Investigations into
      Generation, 1651-1828. Baltimore: Johns Hopkins
      Press, 1966.

      Reviews the ideas about generation that were
prevalent from the mid-seventeenth to the mid-nineteenth
century. Discusses the work of Maupertuis', Wolff,

Haller, Charles Bonnet, Spallanzani, Prevost and Dumas, von Baer, and others.

297. Sandler, Iris. "The Re-examination of Spallanzani's Interpretation of the Role of the Spermatic Animalcules in Fertilization." Journal of the History of Biology 6(1973): 193-223.

Reviews Spallanzani's research on spermatic animalcules in fertilization and explains the basis for the conclusions he drew.

GENETICS

298. Adams, Mark B. "The Founding of Population Genetics: Contributions of the Chetverikov School, 1924-1934." Journal of the History of Biology 1(1968): 23-39.

Calls attention to the importance of studies on the genetic analysis of natural populations (Drosophila) carried out by Sergei S. Chetverikov and his students N.V. Timofeev-Resovsky and N.P. Dubinin.

299. Adams, Mark B. "From 'Gene Fund' to 'Gene Pool': On the Evolution of Evolutionary Language." Studies in the History of Biology 3(1979): 241-285.

Examines the history of the term "gene pool" used in the language of evolution and of population genetics since 1950 when it was introduced by Dobzhansky. Discusses its possible roots in the Russian "gene fund" used from 1926 to 1932.

300. Allen, Garland E. "Opposition to the Mendelian-Chromosome Theory: The Physiological and Developmental Genetics of Richard Goldschmidt." Journal of the History of Biology 7(1974): 49-92.

Discusses Richard Goldschmidt's research on the inheritance and development of sex and on the genetics

of speciation.   Examines his objections to Thomas Hunt
Morgan's Drosophila group's focus on the gene as
transmitter and Goldschmidt's alternate theories which
focused on gene function, on physiological and
developmental questions.

301.   Allen, Garland E.   "T.H. Morgan and the Emergence
        of a New American Biology."   Quarterly Review of
        Biology 44(1969): 168-188.

        Describes the early twentieth century focus on
descriptive morphology in biology and the change to an
experimental focus, using Thomas Hunt Morgan's efforts
to effect the change as an example.

302.   Baxter, Alice Levine.   "Edmund B. Wilson as a
        Preformationist.   Some Reasons for His
        Acceptance of the Chromosome Theory."   Journal
        of the History of Biology 9(1976): 29-57.

        Establishes Edmund Beecher Wilson's position as a
preformationist.   Explains Wilson's variant definitions
of epigenesis and preformation and his theory of
development.   Analyzes Wilson's acceptance of the
chromosome theory and the basis for that acceptance.

303.   Bennet, J.H.   Experiments in Plant Hybridisation.
        Edinburgh: Oliver and Boyd, 1965.

        Contains "Mendel's Original Papers in English
Translation with Commentary and Assessment by the Late
Sir Ronald A. Fisher Together with a Reprint of W.
Bateson's Biographical Notice of Mendel."

304.   Brush, Stephen G.   "Nettie M. Stevens and the
        Discovery of Sex Determination by Chromosomes."
        Isis 69(1978): 163-172.

        Clarifies the major role of Nettie Maria Stevens
in the discovery of sex determination by chromosomes.

305.   Buican, Denis.   Histoire de la Genetique et de
        l'Evolutionisme en France.   Paris: Presses
        Universitaires de France, 1984.

Surveys the history of the French controversies over genetics and evolution.

306. Carlson, Elof Axel. "The Drosophila Group: The Transition from the Mendelian Unit to the Individual Gene." Journal of the History of Biology 7(1974): 31-48.

Explores the scientific and personal interactions of the Drosophila group, Thomas Hunt Morgan, Alfred Henry Sturtevant, Hermann Joseph Muller, and Calvin Blackman Bridges during their research on the gene, primarily between 1910 and 1915.

307. Carlson, Elof Axel. The Gene: A Critical History. Philadelphia: W.B. Saunders, 1966.

Covers Mendel's rediscovery to mid-1960s for students of genetics and history of science. Discusses in detail the research and the main themes of the development of the gene concept.

308. Carlson, Elof Axel. Genes, Radiation, and Society: The Life and Work of H.J. Muller. Ithaca: Cornell University Press, 1981.

Discusses Hermann Joseph Muller's research on genetics and on x-ray induced mutations. Covers his eugenics and political beliefs as well, and the roles they played in shaping his career.

309. Churchill, Frederick B. "William Johannsen and the Genotype Concept." Journal of the History of Biology 7(1974): 5-30.

Discusses Wilhelm Johannsen's pure line research, his original vertical phenotype-genotype distinction, and the influences which brought about its later modification to the horizontal distinction that was so important to research on the gene after 1910.

310. Cock, A.G. "William Bateson's Rejection and Eventual Acceptance of Chromosome Theory." Annals of Science 40(1983): 19-59.

Discusses previously given reasons for William Bateson's positions on the chromosome theory. Proposes several different reasons. Compares and contrasts Bateson's position with that of Wilhelm Johannsen.

311. Dunn, L.C. A Short History of Genetics: The Development of Some of the Main Lines of Thought: 1864-1939. New York: McGraw-Hill, 1965.

Extends Dunn's 1964 Hideyo Noguchi Lectures presenting the major themes of genetics. Covers pre-Mendelian genetics; emphasizes Mendel's work and the research of the decade following it; treats human and population genetics.

312. Gilbert, Scott F. "The Embryological Origins of the Gene Theory." Journal of the History of Biology 11(1978): 307- 351.

Discusses the roots of gene theory in embryological research, particularly that of Edmund Beecher Wilson and Thomas Hunt Morgan. Proposes their differing positions on the chromosome are a result of their earlier embryological research and theories. Describes clearly the question of the site of control of heredity and development and research on the mechanism of sex determination.

313. Glass, Bentley. "The Long Neglect of a Scientific Discovery: Mendel's Laws of Inheritance." In Studies in Intellectual History, edited by George Boas, 148-160. Baltimore: Johns Hopkins Press, 1953.

Discusses the criterion of prematurity as defined by Stent in the cases of Gregor Mendel, Friedrich Miescher, Sir Archibald Garrod and of Oswald T. Avery, C.M. MacLeod, and Maclyn McCarty.

314. Glass, Bentley. "The Long Neglect of Genetic Discoveries and the Criterion of Prematurity." Journal of the History of Biology 7(1974): 101-110.

Treats the discoveries of Gregor Mendel, Friedrich Miescher, and A.E. Garrod as exemplars of G.S. Stent's concept of "prematurity" and argues that the work of Oswald T. Avery, C.M. MacLeod, and Maclyn McCarty is not. Suggests that "lack of generality" is also a relevant criterion.

315. Iltis, Hugo. Life of Mendel. Translated by Eden and Cedar Paul. London: Allen & Unwin, 1932.

Remains a standard introduction to the life and work of Gregor Mendel.

316. Jacob, Francois. The Logic of Life; A History of Heredity. Translated by Betty E. Spillman. New York: Pantheon Books, 1973.

Analyzes the development of ideas on the nature of heredity and reproduction from the sixteenth to the twentieth century.

317. Kevles, Daniel J. "Genetics in the United States and Great Britain, 1890-1930: A Review with Speculations." Isis 71(1980): 441-455.

Discusses major historical publications of the previous fifteen years on the history of genetics, its concepts, its emergence as a discipline, and its controversies. Suggests areas for further research.

318. Kevles, Daniel J. "Genetics in the United States and Great Britain, 1890-1930: A Review with Speculations." In Biology, Medicine and Society 1840-1940, edited by Charles Webster, 193-215. Cambridge: Cambridge University Press, 1981.

Reviews the heated controversy between the biometricians and the Mendelians in England and its historiography. Compares the situation with genetics research in the United States; suggests that institutional factors in the two countries played an important role in the differing responses.

319.   Klein, Aaron E. Threads of Life: Genetics from
       Aristotle to DNA. New York: American Museum of
       Natural History, 1970.

       Contains brief popularized introduction to the
history of genetics.

320.   Krizenecky, Jaroslav, ed. Fundamenta Genetica.
       Brno: Moravian Museum; Prague: Publishing House
       of the Czechoslovak Academy of Science, 1965.

       Contains the revised edition of Gregor Mendel's
classic paper. Includes reprints of twenty-seven
classic papers published from 1899 through 1903,
including those by Hugo de Vries, Carl Correns, and E.
Tschermak.

321.   Ludmerer, Kenneth M. "American Geneticists and
       the Eugenic Movement: 1905-1935." Journal of
       the History of Biology 2(1969): 337-362.

       Analyzes the attitudes of American geneticists
towards eugenics between 1905 and 1935. Explains the
differing roles of internal and external factors at
various stages in the transition from support to
repudiation of the eugenics movement.

322.   Ludmerer, Kenneth M. Genetics and American
       Society: A Historical Appraisal. Baltimore:
       Johns Hopkins University Press, 1972.

       Discusses the eugenicists use of genetics, the
geneticists eventual repudiation of eugenics, the
negative effect eugenics had on the field of genetics,
and the regrowth of genetics after World War II.

323.   MacKenzie, Donald. "Sociobiologies in
       Competition: The Biometrician-Mendelian Debate."
       In Biology, Medicine and Society 1840-1940,
       edited by Charles Webster, 243-288. Cambridge:
       Cambridge University Press, 1981.

       Interprets the turn of the century clash between
the Mendelian geneticists and the biometricians in
sociobiological terms. Examines the social structure of

science (such as differing concepts of biology and the appropriate training and competencies needed to practice it, of competition in its field, and of evolution) and of society at large (such as the eugenics movement).

324. Norton, Bernard J. "Biology and Philosophy: the Methodological Foundations of Biometry." Journal of the History of Biology 8(1975): 85-93.

Suggests Karl Pearson's methodological and ontological (against the uses of homogenous classes) positions as another possible explanation of his opposition to Mendelian genetics.

325. Paul, Diane B. "A War on Two Fronts: J.B.S. Haldane and the Response to Lysenkoism in Britain." Journal of the History of Biology 16(1983): 1-37.

Describes the responses to Lysenkoism of British biologists. Uses J.B.S. Haldane as an exemplar and discusses the conflict between his scientific beliefs and his loyalty to the communist party when its scientific line differed.

326. Provine, William B. "Francis B. Sumner and the Evolutionary Synthesis." Studies in the History of Biology 3(1979): 211-240.

Analyzes the field and laboratory research on the deer mouse of Francis Bertody Sumner. Discusses the significant role he played in the evolutionary synthesis as the first biologist to combine work with natural populations and experimental laboratory work.

327. Ravin, Arnold W. The Evolution of Genetics. New York: Academic, 1965.

Provides a general introductory historical survey of genetics since the 1940s for novices to the field.

328. Ravin, Arnold W. "The Gene as Catalyst; The Gene
      as Organism." Studies in History of Biology
      1(1977): 1-45.

      Traces the development of a distinction between
the transmission of reproducible material and its
developmental expression. Discusses the role of
catalysis and of Leonard Troland's creation of a
physicochemical model.

329. Roll-Hansen, Nils. "Drosophila Genetics: A
      Reductionist Research Program." Journal of the
      History of Biology 11(1978): 159-210.

      Describes the Drosophila research program as a
reductionist one and discusses the contributions of E.B.
Wilson, Thomas Hunt Morgan, and H.J. Muller to that
program. Reviews the empiricist criticisms of William
Bateson and Wilhelm Johannsen.

330. Rosenberg, Charles E. "The Social Environment of
      Scientific Innovation: Factors in the
      Development of Genetics in the U.S." In No
      Other Gods, edited by Charles E. Rosenberg, 196-
      209. Baltimore: Johns Hopkins Press, 1976.

      Suggests reasons why the United States became a
leader in the field of genetics after 1900, including
established interest in heredity, plant and animal
breeding, and academic biology.

331. Shine, Ian, and Sylvia Wrobel. Thomas Hunt
      Morgan: Pioneer of Genetics. Lexington:
      University Press of Kentucky, 1976.

      Describes briefly Thomas Hunt Morgan's life and
Drosophila research in a more popularized style.

332. Stern, Curt, and Eva R. Sherwood, eds. The Origin
      of Genetics: A Mendel Source Book. San
      Francisco: Freeman, 1966.

      Reprints in translation papers by Gregor Mendel,
Wilhelm Olbers Focke, Hugo de Vries, Carl Correns, and
letters from Mendel to Carl Nägeli, and letters from de

Vries and Correns to H.F. Roberts. Includes R.A. Fisher's 1936 article "Has Mendel's Work Been Rediscovered?" and a brief chapter by Sewall Wright.

333. Stubbe, Hans. A History of Genetics from Prehistoric Times to the Rediscovery of Mendel's Laws. Translated by T.R.W. Waters. Cambridge: M.I.T. Press, 1972.

Surveys the major themes in the history of reproduction and heredity; remains a standard.

334. Sturtevant, Alfred Henry. "The Early Mendelians." Proceedings of the American Philosophical Society 109(1965): 199-204.

Identifies and describes briefly the researchers who began working on Mendelian genetics in 1900; characterizes them by age (mostly young) and comments on the fields in which they were working: three plant physiologists, five experimental embryologists, three statisticians, three cytologists, etc.

335. Sturtevant, Alfred Henry. A History of Genetics. New York: Harper & Row, 1965.

Provides an overview of the history of genetics from a biographical point of view. Covers up to Mendel's rediscovery in three chapters, with the major portion of the book covering twentieth century developments. Includes a chronology and diagrams of teacher-student relationships.

336. Taylor, J. Herbert, ed. Selected Papers on Molecular Genetics. New York: Academic Press, 1965.

Includes reprints of significant papers on biochemical genetics, the nature of genetic material, DNA structure and replication, genetic recombination, and the function of genetic material, with brief historical introductions to the five sections.

337.   Veer, P.H.W.A. de. Leven en Werk van Hugo de
        Vries. Groningen, Netherlands: Wolters-
        Noordhoff, 1969.

        Discusses Hugo de Vries research in genetics and
cell physiology.

338.   Voeller, Bruce R., ed. The Chromosome Theory of
        Inheritance; Classic Papers in Development and
        Heredity. New York: Appleton, 1968.

        Collects twenty-five significant papers
(translated where necessary) on the chromosome theory of
inheritance from the nineteenth and twentieth centuries,
arranged under the topics: "Role of the Nucleus in
Fertilization," "Role of the Nucleus as a Vehicle for
Inheritance," "Equivalence of Contribution of the Two
Parents," "Continuity of Chromosomes," "Individuality of
Chromosomes," "Mendel, Mendelian Genetics and the
Chromosome Theory of Inheritance;" each article is
accompanied by editorial comment.

339.   Wagner, Robert P., ed. Genes and Proteins.
        Stroudsburg, Pennsylvania: Dowden, Hutchinson
        and Ross, 1975.

        Collects reprints of thirty-six classic papers in
genetics. Includes brief introductions to papers.

340.   Whitehouse, H.L.K. The Mechanism of Heredity.
        London: Arnold, 1965.

        Summarizes theories of hereditary mechanisms from
the direct inheritance of characters through the operon.
Presents the research which confirmed or disconfirmed
each theory.

HEREDITY

341.   Allen, Garland E. "Hugo de Vries and the
        Reception of the Mutation Theory." Journal of
        the History of Biology 2(1969): 55-87.

        Discusses Hugo de Vries' mutation theory and the

reasons why it was so popular between 1900 and 1910.
Cites the misunderstanding of natural selection and of
the nature of species, along with the desire for
testable, quantitative, analytical/reductive
explanations as reasons.

342. Allen, Garland E. "The Introduction of Drosophila
into the Study of Heredity and Evolution, 1900-
1910." Isis 66(1975): 322-333.

Demonstrates the widespread use of Drosophila
prior to 1910 and details how Thomas Hunt Morgan came to
be familiar with and use Drosophila for experiments in
heredity.

343. Blacher, Leonidas I. The Problem of the
Inheritance of Acquired Characters: A History
of A Priori and Empirical Methods Used to Find a
Solution. English translation edited by
Frederick B. Churchill. Washington, D.C.:
Smithsonian Institution Libraries, 1982.

Covers the controversy over the inheritance of
acquired characteristics from Lamarck through the mid-
twentieth century, discussing the ideas and research of
Lamarck, Charles Darwin, Ernst Haeckel, and Paul
Kammerer, among others. Includes chapters on
paleontologists' views and on the Soviets' views of the
problem, and a chapter on the inheritance of acquired
behavioral traits.

344. Bowler, Peter J. "E.W. MacBride's Lamarckian
Eugenics and Its Implications for the Social
Construction of Scientific Knowledge." Annals
of Science 41(1984): 245-260.

Describes Ernest William MacBride's uses of
Lamarckism in support of his belief in eugenics.

345. Burkhardt, Richard W., Jr. "Closing the Door on
Lord Morton's Mare: The Rise and Fall of
Telegony." Studies in History of Biology
3(1979): 1-21.

Recounts the story first reported in 1820 to the

Royal Society of the effect of the mating of Lord
Morton's mare and the quagga on the mare's later
offspring.   Analyzes the case in terms of early
hereditary theory and of dealing with anomalous
evidence.

346. Burnham, John C.   "Instinct Theory and the German
      Reaction to Weismannism."   Journal of the
      History of Biology 5(1972):   321-326.

      Compares and contrasts the German and the Anglo-
American reactions to August Weismann's germ plasm
theory, in particular the concept of instincts as
species specific traits.   Focuses on the work of
Heinrich Ernst Ziegler and his reductionist definition
of instinct.

347. Campbell, Margaret.   "The Concepts of Dormancy,
      Latency, and Dominance in Nineteenth-Century
      Biology."   Journal of the History of Biology
      16(1983): 409-431.

      Explains clearly the meanings of Gregor Mendel's
dominance, Charles Darwin's dormancy and Hugo de Vries'
latency and firmly differentiates each one from the
other.

348. Campbell, Margaret.   "Did de Vries Discover the
      Law of Segregation Independently?"   Annals of
      Science 37(1980): 639-655.

      Builds a case that Hugo de Vries did not
discover the law of segregation independently of Gregor
Mendel.

349. Churchill, Frederick B.   "August Weismann and a
      Break from Tradition."   Journal of the History
      of Biology 1(1968): 91-112.

      Traces August Weismann's switch from the
concepts of Ernst Haeckel and the development of
Weismann's germ-plasm theory.

350.   Churchill, Frederick B.   "Hertwig, Weismann, and
the Meaning of Reduction Division circa 1890."
Isis 61(1970): 429-457.

Explains and compares the theories of Oscar
Hertwig and August Weismann and the cytological evidence
supporting them both.

351.   Cock, A.G.   "William Bateson, Mendelism and
Biometry." Journal of the History of Biology
6(1973): 1-36.

Discusses William Bateson's research on variation
and heredity pre and post Mendel. Reviews his dispute
with the biometricians, Sir Ronald Fisher in particular,
and the inductive and hypothetico-deductive
characteristics of Bateson and his opponents.

352.   Coleman, William.   "Limits of the Recapitulation
Theory: Carl Friedrich Kielmeyer's Critique of
the Presumed Parallelism of Earth History,
Ontogeny, and the Present Order of Organisms."
Isis 64(1973): 341-350.

Bases the discussion on a manuscript letter of
November 25, 1804 from Carl Friedrich Kielmeyer to Karl
Joseph Hieronymous Windischmann.

353.   Cowan, Ruth Schwartz.   "Francis Galton's
Contribution to Genetics." Journal of the
History of Biology 5(1972): 389-412.

Demonstrates that Francis Galton's contribution
to genetics was his redefinition of heredity as a
researchable relationship incorporating variation and
reversion.

354.   Cowan, Ruth Schwartz.   "Francis Galton's
Statistical Ideas: The Influence of Genetics."
Isis 63(1972): 509-528.

Discusses Galton's research and discovery of
regression and correlation. Demonstrates its basis in
his studies of heredity for the purpose of eugenics.

355. Cowan, Ruth Schwartz. "Nature and Nurture: The
      Interplay of Biology and Politics in the Work of
      Francis Galton." Studies in the History of
      Biology 1(1977): 133-208.

      Discusses Galton's scientific contributions to
genetics, statistics, psychology, and anthropology and
their basis in his social and political commitment to a
eugenic society.

356. Cowan, Ruth Schwartz. Sir Francis Galton and the
      Study of Heredity in the Nineteenth Century.
      New York: Garland, 1985.

      Analyzes Francis Galton's contributions to
genetics (heredity defined as a researchable
relationship incorporating both variation and
reversion), statistics (regression and correlation), and
psychology (anthropomorphic method). Discusses their
interrelatedness based on Galton's belief in the
significance of nature over nurture as a component of
his commitment to eugenics.

357. Cravens, Hamilton. The Triumph of Evolution:
      American Scientists and the Heredity-Environment
      Controversy 1900-1941. Philadelphia: University
      of Pennsylvania Press, 1978.

      Analyzes nature-nurture controversies, 1900-
1941, from biological and social standpoints. Includes
chapters on biology, psychology, sociology, on eugenics,
race, and evolution, on instinct and on mental testing.

358. Darden, L. "William Bateson and the Promise of
      Mendelism." Journal of the History of Biology
      10(1977): 87-106.

      Explains William Bateson's reasons for
recognizing the promise of Mendelism and generalizes
from them elements for the recognition of promise in
other areas of science.

359. Gaissinovitch, A.E. "Problems of Variation and
      Heredity in Russian Biology in the Late

Nineteenth Century." Journal of the History of Biology 6(1973): 97-123.

Discusses research on the influence of environment in changing species characteristics and on the inheritance of acquired characteristics done in the 1870s and 1880s by such biologists as V.I. Shmankevitch, I.I. Mechnikov, K.A. Timiryazev, and P.F. Leshaft, who were debating various theories of variation, heredity, and evolution. Contrasts these with biologists in the 1890s, such as N.A. Kholodkovsky, M.A. Menzbier, and V.M. Shemkevitch, who were interested in August Weismann's theory.

360. Gasking, Elizabeth B. "Why Was Mendel's Work Ignored?" Journal of the History of Ideas 20(1959): 60-84.

Suggests that Gregor Mendel's work went unrecognized because both applied and theoretical biologists were focused on different issues, the horticulturists on creating new species in wild plants and the theoreticians, as a result of Darwin's work, on finding historical evidence of evolution in plants and animals or on embryology.

361. Koestler, Arthur. The Case of the Midwife Toad. London: Hutchinson, 1971.

Presents a biography of Paul Kammerer and a lively partisan description of the controversy surrounding his research on the nuptial pads of the midwife toad and the inheritance of acquired characters.

362. Maienschein, Jane. "Cell Lineage, Ancestral Reminiscence, and the Biogenetic Law." Journal of the History of Biology 11(1978): 129-158.

Discusses the contributions of the American embryologists Charles Otis Whitman, Edmund Beecher Wilson, Edwin Grant Conklin, Aaron Louis Treadwell, Albert Davis Mead, and Frank Rattray Lillie to the understanding of ontogeny and its relation to phylogeny.

363.   Olby, Robert C. Origins of Mendelism. London:
        Constable, 1966.

        Surveys in a popularized style pre-Mendelian
concepts of reproduction, practices of hybridization,
and theories of heredity; describes Gregor Mendel's work
and its rediscovery.

364.   Orel, Vitezlav, and Roger Wood. "Early
        Developments in Artificial Selection As a
        Background to Mendel's Research." History and
        Philosophy of the Life Sciences 3(1981): 145-
        170.

        Outlines the development of artificial breeding
of stock, sheep in particular, in eighteenth and
nineteenth century Europe.

365.   Orel, Vitezlav, and M. Vavra. "Mendel's Program
        for the Hybridization of Apple Trees." Journal
        of the History of Biology 1(1968): 219-224.

        Describes briefly a program of hybridization of
apple trees based on Mendel's manuscript notes.

366.   Rinard, Ruth G. "The Problem of the Organic
        Individual: Ernst Haeckel and the Development
        of the Biogenetic Law." Journal of the History
        of Biology 14(1981): 249-276.

        Demonstrates the influence of Johannes Müller and
Alexander Braun's ideas on the development of Ernst
Haeckel's biogenic law.

367.   Robinson, Gloria. A Prelude to Genetics:
        Theories of a Material Substance of Heredity:
        Darwin to Weismann. Lawrence, Kansas: Coronado
        Press, 1979.

        Discusses nineteenth century attempts to account
for inheritance, covering the theories and research of
Charles Darwin, Francis Galton, Ernst Haeckel, Gustav
Jaeger, William Keith Brooks, Carl von Nägeli, August
Weismann, and Hugo de Vries.

368. Roger, Jacques. <u>Les Sciences de la Vie dans la</u>
      <u>Pensée Française du XVIIIe Siècle: La Génération</u>
      <u>des Animaux de Descartes a l'Encyclopédie.</u>
      Paris: Colin, 1972.

      Encyclopedic discussion of generation,
reproduction, and heredity.

369. Roll-Hansen, Nils. "The Genotype Theory of
      Wilhelm Johannsen and Its Relation to Plant
      Breeding and the Study of Evolution."
      <u>Centaurus</u> 22(1978): 201-235.

      Explains in detail the genotype theory of Wilhelm
Johannsen, emphasizing that his genotype was an
Aristotelian biological ideal form and that Johannsen's
gene was not a separate part of a cell.

370. Russell, E.S. <u>The Interpretation of Development</u>
      <u>and Heredity: A Study in Biological Method.</u>
      Oxford: Clarendon Press, 1930 or reprint 1972,
      Freeport, N.Y., Books for Libraries Press.

      Remains a standard early source for historical
discussion of issues of heredity and development,
including theories of preformation, epigenesis, germ-
plasm, and the cell.

371. Sandler, Iris. "Pierre Louis Moreau de Maupertuis
      —A Precursor of Mendel?" <u>Journal of the History</u>
      <u>of Biology</u> 16(1983): 101-136.

      Distinguishes between Gregor Mendel and Pierre
Louis Moreau de Maupertuis' interpretations of
dominance, segregation, and the laws of probability.
Concludes that Maupertuis was not a precursor of Mendel.

372. Sapp, Jan. "The Struggle for Authority in the
      Field of Heredity, 1900-1932: New Perspectives
      on the Rise of Genetics." <u>Journal of the</u>
      <u>History of Biology</u> 16(1983): 311-342.

      Describes the field of heredity and the
disciplines working on it at the beginning of this
century. Discusses how genetics became the dominant

discipline in the study of heredity. Suggests that the
genotype-phenotype distinction served to exclude other
disciplines.

373. Weinstein, Alexander. "How Unknown Was Mendel's
      Paper?" Journal of the History of Biology
      10(1977): 341-364.

Identifies a citation of Gregor Mendel's
"Versuche uber Pflanzen-Hybrides" that appeared in 1881
in Guide to the Literature of Botany by Benjamin Daydon
Jackson and discusses where Jackson obtained the
citation and whether he read it or understood its
significance.

374. Weir, J.A. "Agassiz, Mendel, and Heredity."
      Journal of the History of Biology 1(1968): 179-
      203.

Reproduces, with commentary, Louis Agassiz'
lecture on animal breeding at a public meeting of the
Massachusetts State Board of Agriculture at Greenfield
on December 15, 1864. Relates these views of
inheritance to the work of Gregor Mendel.

375. Wells, Kentwood D. "Sir William Lawrence (1783-
      1867): A Study of Pre-Darwinian Ideas on
      Heredity and Variation." Journal of the History
      of Biology 4(1971): 319-361.

Clarifies Sir William Lawrence's positions on
heredity and race formation in man and on evolution as
treated in his Lectures on Physiology, Zoology, and the
Natural History of Man.

376. Zirkle, Conway. "The Early History of the Idea of
      the Inheritance of Acquired Characters and of
      Pangenesis." Transactions of the American
      Philosophical Society 35(1946): 91-151.

Presents quotations (in translation where
necessary) from the classic accounts of the inheritance
of acquired characters from the sixteenth-nineteenth
centuries and of pangenesis from the thirteenth-
nineteenth centuries.

377. Zirkle, Conway. "Gregor Mendel & His Precursors."
      Isis 42(1951): 97-104.

      Identifies findings relevant to the research of
Gregor Mendel in studies of eighteenth and nineteenth
century plant and bee hybridizers particularly C.F. v.
Gartner and Johann Dzierzon.

378. Zirkle, Conway. "The Role of Liberty Hyde Bailey
      and Hugo de Vries in the Rediscovery of
      Mendelism." Journal of the History of Biology
      1(1968): 205-218.

      Recounts versions of Hugo de Vries' learning of
Gregor Mendel's article. Dates de Vries reading of
Mendel's paper during the six months after his
presentation at the Hybrid Conference in July 1899 and
his receipt of the page proofs of the Conference
proceedings. Identifies Liberty Hyde Bailey's 1892
bibliography in The Rural Library and/or a reprint of
Mendel's article sent by M.W. Beyerinck as de Vries'
possible sources of the Mendel citation.

HISTOLOGY

379. Aterman, K. "Connective Tissue: An Eclectic
      Historical Review with Particular Reference to
      the Liver." Histochemical Journal 13(1981):
      341-396.

      Surveys the development of knowledge about
connective tissue of the liver.

380. Aterman, K. "The Development of the Concept of
      Lysosomes. A Historical Survey, with
      Particular Reference to the Liver."
      Histochemical Journal 11(1979): 503-541.

      Reviews the major events in the development of
understanding of lysosomes, including the development of
staining.

381. Clark, George, and Frederick H. Kasten, eds.
      History of Staining. 3d ed. Baltimore:

Williams & Wilkins, 1983.

Reprints articles from Stain Technology 1928 to
1933 and adds a biography of Ralph Dougall Lillie and
chapters on the history of connective tissue stains,
neurological stains, immunological staining,
fluorescence microscopy and staining, and histochemistry
of proteins and nucleic acids.

382. Jacyna, L.S. "John Goodsir and the Making of
      Cellular Reality." Journal of the History of
      Biology 16(1983): 75-99.

Describes the difficulties in establishing a
model by which understanding of new information
presented by use of the achromatic microscope could be
achieved. Discusses John Goodsir's use of the
"transcendental" anatomy of gross structure as a model
in his teaching and writing and how that affected his
physiological concepts and his student's careers.

LIMNOLOGY

383. Frey, David G. "Wisconsin: the Birge-Juday Era."
      In Limnology in North America, edited by David
      G. Frey, 3-54. Madison: University of Wisconsin
      Press, 1963.

Reviews the limnological research of E.A. Birge
and Chancey Juday conducted 1875-1944 in Wisconsin.

384. Lauff, George H. "A History of the American
      Society of Limnology and Oceanography." In
      Limnology in North America, edited by David G.
      Frey, 667-682. Madison: University of
      Wisconsin, 1963.

Recounts the history of the society from its
founding in 1936, including its committees, programs,
and publications.

MARINE BIOLOGY AND OCEANOGRAPHY

385. Bronk, Detlev W. "Marine Biological Laboratory:
      Origin and Patrons." Science 189(1975): 613-
      617.

Outlines the early development of the physical
plant and the funding base of the Marine Biological
Laboratory at Woods Hole.

386. Burstyn, Harold L. "Reviving American
      Oceanography: Frank Lillie, Wickliffe Rose, and
      the Founding of the Woods Hole Oceanographic
      Institution." In Oceanography: The Past, edited
      by Mary Sears and Daniel Merriman, 57-66. New
      York: Springer-Verlag, 1980.

Outlines briefly events in the mid-twenties that
lead to the establishment of support for and the
incorporation of the Woods Hole Oceanographic
Institution to study marine biology and physical
oceanography.

387. Colin, Patrick L. "A Brief History of the
      Tortugas Marine Laboratory and the Department of
      Marine Biology, Carnegie Institution of
      Washington." In Oceanography: The Past, edited
      by Mary Sears and Daniel Merriman, 138-147. New
      York: Springer-Verlag, 1980.

Covers briefly two decades of the Tortugas Marine
Laboratory under Alfred G. Mayor. Identifies scientists
and research conducted there from its establishment in
1902.

388. Colloque International sur l'Histoire de la
      Biologie Marine: Les Grandes Expéditions
      Scientifiques et la Création des Laboratoires
      Maritimes. Paris: Masson, 1965.

Contains twenty-eight papers presented at a
conference held in 1963; covers a broad range of topics
including those not frequently treated, such as Chinese
marine biology, along with the more traditional, such as
William Coleman's "Les organisms marines et l'anatomie

comparée dite expérimentale: L'oeuvre de Georges
Cuvier."

389.   Deacon, Margaret B., ed. Oceanography: Concepts
        and History. Stroudsburg, Pa.: Dowden,
        Hutchinson & Ross, 1978.

        Contains thirty-nine articles on oceanography, of
which three are classic contributions to nineteenth
century marine biology preceded by brief historical
comment.

390.   Dexter, Ralph W. "The Annisquam Sea-side
        Laboratory of Alpheus Hyatt, Predecessor of the
        Marine Biological Laboratory at Woods Hole,
        1880-1886." In: Oceanography: The Past, edited
        by Mary Sears and Daniel Merriman, pp. 94-100.
        New York: Springer-Verlag, 1980.

        Outlines briefly the establishment in 1880 of a
summer sea-side laboratory at Annisquam, Massachusetts
and its move in 1898 to Woods Hole.

391.   Idyll, Clarence P., ed. Exploring the Ocean
        World: A History of Oceanography. New York:
        Crowell, 1969.

        Surveys in a readable popular style a century of
oceanography, including the history of physical,
chemical, biological, and geological oceanography
written by experts in the fields.

392.   Lillie, Frank R. The Woods Hole Marine Biological
        Laboratory. Chicago: University of Chicago
        Press, 1944.

        Covers the development of marine biological
laboratories briefly and of the Woods Hole Laboratory in
detail, including facilities, staff, administration,
funding, and research carried out there.

393.   Oppenheimer, Jane M. "Some Historical Backgrounds
        for the Establishment of the Stazione Zoologica
        at Naples." In Oceanography: The Past, edited

by Mary Sears and Daniel Merriman, 179-187. New
York: Springer-Verlag, 1980.

Considers factors that led to interest in marine
biological stations and to the establishment of the
first successful one.

394.   Raitt, Helen, and Beatrice Moulton. Scripps
        Institution of Oceanography: First Fifty Years.
        Los Angeles: Ward Ritchie Press, 1967.

Covers chronologically, including all the
important names, dates, events, and issues in the origin
and development of Scripps from William Emerson Ritter's
first efforts in 1892.

395.   Rehbock, Philip F. "The Early Dredgers:
        'Naturalizing' in British Seas, 1830-1850."
        Journal of the History of Biology 12(1979): 293-
        368.

Describes the marine zoology dredging of John
Forbes and other naturalists, and the activities of the
Dredging Committee of Section D of the British
Association, and discusses the issues with which they
were concerned, such as the distribution in space and
time of the specimens they collected.

396.   Rehbock, Philip F. "Huxley, Haeckel, and the
        Oceanographers: The Case of Bathybius
        Haeckelii." Isis 66(1975): 504-533.

Describes the controversy over the Bathybius and
the roles of Thomas Henry Huxley and Ernst Haeckel.
Analyzes it as an error of interpretation and an error
congruent with its time.

397.   Schlee, Susan. The Edge of an Unfamiliar World: A
        History of Oceanography. New York: Dutton,
        1973.

Discusses oceanography from its inception in the
mid-nineteenth century with good coverage of natural
history and marine biology, especially American and
British.

398. Schlee, Susan. On Almost Any Wind: The Saga of
      the Oceanographic Research Vessel Atlantis.
      Ithaca, N.Y.: Cornell University Press, 1978.

      Covers the adventures of the Atlantis, the
research vessel of the Woods Hole Oceanographic
Institution from 1931 to 1966. Narration includes
popularized description of the marine biological
research carried out from the ship.

399. Schlee, Susan. "The R/V Atlantis and Her First
      Oceanographic Institution." In Oceanography:
      The Past, edited by Mary Sears and Daniel
      Merriman, 49-56. New York: Springer, 1980.

      Emphasizes the influence the research vessel
Atlantis had on the development of the Woods Hole
Oceanographic Institution.

400. Sears, Mary and Daniel Merriman, eds.
      Oceanography: The Past. New York: Springer-
      Verlag, 1980.

      Contains sixty-nine papers, the proceedings of
the Third International Congress on the History of
Oceanography held Sept. 22-26, 1980.

401. Uschmann, Georg. "Haeckel's Biological
      Materialism." History and Philosophy of the
      Life Sciences 1(1979): 101-118.

      Discusses the philosophical concepts in the
publications of and morphologic and taxonomic research
on the marine invertebrates radiolaria, medusae,
echinoderms, and sponges by Ernst Haeckel.

METABOLIC SYSTEM

402. Galdston, Iago, ed. Human Nutrition: Historic and
      Scientific. New York: International
      Universities Press, 1960.

      Includes historical chapters by Owsei Temkin,
"Nutrition From Classical Antiquity to the Baroque," by

E.W. McHenry, "From Lavoisier to Beaumont and Hopkins," and by Elmer V. McCollum, "From Hopkins to the Present."

403. Goldblith, Samuel A., and Maynard A. Joslyn. Milestones in Nutrition. Westport, Connecticut: AVI Publishing, 1964.

Collects forty-three classic articles relating to the discovery of vitamins. Includes brief commentary and biographical information and begins with a chapter on the history of metabolism (see under Graham Lusk).

404. Holmes, Frederick L. "The Transformation of the Science of Nutrition." Journal of the History of Biology 8(1975): 135-144.

Discusses eighteenth century explanations of nutrition and the difficulties of demonstrating which was useful. Identifies the research of Friedrich Tiedemann and Leopold Gmelin as the emergence of a "field of discourse".

405. Lusk, Graham. "A History of Metabolism." In Endocrinology and Metabolism, edited by Lewellys F. Barker, vol. 3, 1-78. New York: D. Appleton, 1922.

Identifies major scientists and their contributions to nutrition through the nineteenth century.

406. McCay, Clive Maine. Notes on the History of Nutritional Research. Edited by F. Verzar. Berne: Huber, 1973.

Covers research contributions during the nineteenth and twentieth (to 1940) centuries to the field of nutrition. Based on McCay's lecture notes arranged in five areas: "Finding your Way in the Nutrition Literature," "Three Great Problems of Nutrition and Biochemistry," "Proteins and Their Nutritive Value," "Proteins and Pathology," and "Anorganic Substances."

407. McCollum, Elmer Verner. A History of Nutrition.
      Boston: Houghton-Mifflin, 1957.

      Reviews major discoveries in nutritional
investigations from mid-eighteenth to mid-nineteenth
century.

408. O'Hara-May, Jane. "Measuring Man's Needs."
      Journal of the History of Biology 4(1971): 249-
      273.

      Traces the development of Jacob Moleschott's
quantification of dietary needs on the basis of food
intake, body-weight, and time.

409. Robin, Eugene Debs, ed. Claude Bernard and the
      Internal Environment: A Memorial Symposium. New
      York: Dekker, 1979.

      Contains papers presented by scientists at a
centennial symposium in memory of the death of Claude
Bernard. Intends to link past with present, therefore
levels of historical content vary considerably from
paper to paper.

MICROBIOLOGY

410. Brock, Thomas D., trans. and ed. Milestones in
      Microbiology. Englewood Cliffs: Prentice-Hall,
      1961.

      Reprints, in translation where necessary, classic
papers in microbiology, Covers spontaneous generation
and fermentation from Anthony van Leeuwenhoek through
Eduard Buchner and general microbiology from Ferdinand
Cohn through Albert J. Kluyver. Includes a brief
historical introduction.

411. Collard, Patrick. The Development of
      Microbiology. Cambridge: Cambridge University
      Press, 1976.

      Presents the history of the microscope briefly
and covers artificial culture media. Discusses

sterilization, microbial metabolism, microbial genetics, and the classification of bacteria as well as all the medical aspects of microbiology.

412. Doetsch, R.N. Microbiology: Historical
      Contributions From 1776-1908. New Brunswick,
      N.J.: Rutgers University Press, 1960.

     Reprints, translated where necessary, seventeen classic contributions to microbiology, with introductory biographical and explanatory comments for each.

413. Geison, Gerald L. "Pasteur on Vital versus
      Chemical Ferments: A Previously Unpublished
      Paper on the Inversion of Sugar." Isis
      72(1981): 425-445.

     Contains a facsimile and a translation of Pasteur's June 1860 unpublished paper on the inversion of sugar. Discusses Pasteur's ideas about fermentation and his commitment to a biological theory of fermentation.

414. Lechevalier, Hubert A., and Morris Solotorovsky.
      Three Centuries of Microbiology. [revised
      paperback edition] New York: Dover, 1974.

     Surveys the main lines of development of the field of microbiology from a biographical viewpoint and in a popular style. Includes translations of portions of classic papers. Contains chapters on Pasteur, Koch, bacteria, cellular and humoral immunology, soil microbiology, viruses and rickettsiae, mycology, protozoology, chemotherapy, and genetics.

415. Vandervliet, Glenn. Microbiology and the
      Spontaneous Generation Debate during the 1870's.
      Lawrence, Kansas: Coronado Press, 1971.

     Surveys in a brief, popular style the men, the research and the controversy in the 1870s over the spontaneous origin of microbial life.

MICROSCOPES AND MICROSCOPY

416. Afzelius, B.A. "Half a Century of Electron
      Microscopy: the Early Years." Ultrastructural
      Pathology 2(1981): 309-311.

      Outlines the invention of the electron
microscope in 1931 by Ernst Ruska. Describes why it did
not come into use by biologists until after 1950.

417. Bracegirdle, Brian. A History of Microtechnique.
      Ithaca: Cornell University Press, 1978.

      Surveys the history of the micro-tome and
histological methods, including the substances and
instruments used; ends with comments on histology and
medical education.

418. Bradbury, Savile. The Evolution of the
      Microscope. Oxford: Pergamon, 1967.

      Surveys the history of the microscope from the
earliest, including compound, achromatic, optical and
electron.

419. Bradbury, Savile, and Gerard L'Estrange Turner,
      eds. Historical Aspects of Microscopy.
      Cambridge: Royal Microscopical Society, 1967.

      Contains six papers presented at an historical
conference held by the Royal Microscopical Society in
1966, including useful contributions by Savile Bradbury
on "The Quality of the Image Produced by the Compound
Microscope: 1700-1840" and T. Mulvey's "The History of
the Electron Microscope."

420. Clay, Reginald S., and Thomas H. Court. The
      History of the Microscope: Compiled from
      Original Instruments and Documents, up to the
      Introduction of the Achromatic Microscope.
      Boston: Longwood Press, 1978.

      Reprints the classic 1932 work which surveys the
history of magnifiers, the invention of the microscope,

and the development of a variety of kinds of
microscopes, such as simple, compound, Marshall,
Culpeper, box, cuff, solar, scroll, and reflecting.
Includes a list of microscope makers.

421. Marton, L.L. Early History of the Electron
      Microscope. San Francisco Press, 1968.

      Narrates briefly the development of the electron
microscope and its introduction into biology; written by
participant.

422. Nuttall, R.H. Microscopes from the Frank
      Collection 1800-1860, Illustrating the
      Development of the Achromatic Instrument.
      Jersey, Channel Islands: A. Frank, 1979.

      Contains photographs and descriptions of
microscopes from the collection of Arthur Frank, with an
historical introduction by R.H. Nuttall.

423. Turner, Gerard L'Estrange. Essays on the History
      of the Microscope. Oxford: Senecio, 1980.

      Reprints twelve articles by Turner on various
aspects of the history of the microscope; emphasizes the
artifacts.

424. Turner, Gerard L'Estrange. "Micrographia
      Historica: The Study of the History of the
      Microscope." Proceedings of the Royal
      Microscope Society 7(1972): 121-149.

      Reviews histories of the microscope and argues
the need to examine and discuss the microscopes
themselves.

425. Warner, J.H. "'Exploring the Inner Labyrinths of
      Creation': Popular Microscopy in Nineteenth-
      Century America." Journal of the History of
      Medicine 37(1982): 7-33.

      Traces the brief period of amateur use of the
microscope in the United States from 1850 to 1890.

Discusses the handbooks, journals, and societies for amateur microscopists and the reasons for the decline of popular microscopy at the end of the century.

MOLECULAR BIOLOGY

426.   Abir-Am, P. "From Biochemistry to Molecular Biology: DNA and the Acculturated Journey of the Critic of Science Erwin Chargaff." History and Philosophy of the Life Sciences 2(1980): 3-60.

Discusses the transition to molecular biology using the research and career of Erwin Chargaff as an exemplar of the older model of biologist and biochemist who resisted the changes and drawing on autobiographical material from Chargaff's Heraclitean Fire.

427.   Cairns, J., James D. Watson, and Gunther S. Stent, eds. Phage and the Origins of Molecular Biology. Cold Spring: Cold Spring Harbor Laboratory of Quantitative Biology, 1966.

Contains papers presented by biologists discussing the development of their areas of research at a Festschrift for Max Delbruck to record the history of the Phage Group from 1945 to 1966.

428.   Carlson, Elof Axel. "An Unacknowledged Founding of Molecular Biology: H.J. Muller's Contributions to Gene Theory, 1910-1936." Journal of the History of Biology 4(1971): 149-170.

Describes H.J. Muller's contributions to the development of molecular biology, especially his ideas on genetic specificity, the special structure of the gene, the difference of the gene from all other biochemical molecules, and gene activities as the basis of life.

429.   Dodson, G., J.P. Glusker, and D. Sayre, ed. Structural Studies on Molecules of Biological Interest. A Volume in Honour of Professor Dorothy Hodgkin. Oxford: Clarendon, 1981.

Includes seven historical and biographical chapters on Dorothy Hodgkin, molecular biophysics at Oxford, X-ray crystallography in the U.S., and cholesteryl iodide by eminent leaders in the field such as Max F. Peruz and Linus Pauling.

430. Judson, Horace Freeland. The Eighth Day of Creation: Makers of the Revolution in Biology. New York: Simon and Schuster, 1979.

Traces the chief discoveries of the first fifty years of molecular biology - the structure and function of DNA, RNA, Protein. Based on eight years of interviews with 111 scientists, it includes technical details clearly explained, biographical information, and discussion of the external influences that affected the field and its participants.

431. Lwoff, Andre, and Agnes Ullmann, eds. Origins of Molecular Biology. A Tribute to Jacques Monod. New York: Academic Press, 1979.

Contains thirty-two chapters on various aspects of the research carried on at the Institut Pasteur under the leadership of Jacques Monod. Written by the participants: colleagues, staff members, and students.

432. Olby, Robert C. "Francis Crick, DNA and the Central Dogma." Daedalus 99(1970): 938-987.

Describes Francis Crick's education and the beginning of his research career, his research on DNA, and the development of molecular biology.

433. Olby, Robert C. "The Significance of the Macromolecules in the Historiography of Molecular Biology." History and Philosophy of the Life Sciences 1(1979): 185-198.

Outlines four stages in the development of molecular biology. Discusses the importance of Hermann Staudinger's macromolecule.

434. Portugal, Franklin H., and Jack S. Cohen. A
      Century of DNA: A History of the Discovery of
      the Structure and Function of the Genetic
      Substance. Cambridge: MIT Press, 1977.

      Discusses the development of knowledge about
deoxyribonucleic acid (DNA) from discovery of nuclein in
1869 by Friedrich Miescher to the genetic code.

435. Watson, James D. The Double Helix. A Personal
      Account of the Discovery of the Structure of
      DNA. A new critical edition including text,
      commentary, reviews, original papers edited
      by Gunther S. Stent. London: Weidenfeld and
      Nicolson, 1981.

      Includes a reprint of James D. Watson's
autobiographical recounting of the discovery of the
molecular structure of DNA, along with reprints of six
papers by Watson & Francis Crick, Maurice Wilkins, et
al. and Rosalind Franklin & Gosling, accompanying these
are thirteen book reviews and Stent's discussion of
them, as well as Stent's Introduction and chapter on
Watson's first publication of the Double Helix, and
brief perspectives by other participants.

MYCOLOGY

436. Ainsworth, Geoffrey C. Introduction to the
      History of Mycology. Cambridge: Cambridge
      University Press, 1976.

      Outlines the history of major areas of
mycological study, such as taxonomy, reproduction,
genetics, distribution, and classification. Includes
illustrations, chronology, and bibliography.

NATURAL HISTORY

437. Adams, Alexander B. Eternal Quest: The Story of
      the Great Naturalists. New York: G.P. Putnam's,
      1969.

      Surveys significant eighteenth and nineteenth

century contributions to natural history in a popular
style from a biographical viewpoint.

438.   Allen, David Elliston.   The Naturalist in Britain:
       A Social History. London: Allen Lane, 1978.

       Describes and analyzes the growth of the British
natural history movement in relation to society from the
end of the seventeenth through the early twentieth
century.

439.   Beddall, Barbara G.   "The Isolated Spanish Genius
       - Myth or Reality? Felix de Azara and the Birds
       of Paraguay." Journal of the History of Biology
       16(1983): 225-258.

       Argues that Azara's isolation was real and that
his contributions to zoology were important ones.

440.   Beddall, Barbara G.   "'Un Naturalista Original':
       Don Felix de Azara, 1746-1821." Journal of the
       History of Biology 8(1975): 15-66.

       Describes the natural history contributions,
particularly zoological contributions, of don Felix de
Azara, a Spanish military engineer who spent twenty
years in South America.

441.   Broberg, Gunnar, ed. Linnaeus: Progress and
       Prospects in Linnaean Research. Stockholm:
       Almquist & Wiksell, 1980.

       Contains eighteen papers presented at a
bicentenary symposium on Carl Linnaeus. Includes papers
on his zoology and botany, on his German pupils, his
reception in France, and his zoological dissertations.

442.   Buckman, Thomas R., ed. Bibliography & Natural
       History. Lawrence: University of Kansas
       Libraries, 1966.

       Publishes nine papers presented at a 1964
conference on bibliography and natural history.
Includes Frans A. Stafleu on "Redouté and His Circle,"

John C. Greene on the "Founding of the Peale's Museum,"
Jerry Stannard on "Early American Botany and Its
Sources," and Sten Lindroth on "Two Centuries of
Linnaean Studies."

443.   Coleman, William, ed. and trans. The
       Interpretation of Animal Form; Essays by
       Jeffries Wyman, Carl Gegenbauer, E. Roy
       Lankester, Henri Lacaze Duthiers, Wilhelm His
       and H. Newell Martin, 1868-1888. New York and
       London: Johnson Reprint Corporation, 1967.

       Reprints classic essays on symmetry and homology
in limbs, the condition and significance of morphology,
degeneration, the study of zoology, the principles of
animal morphology, and the study and teaching of
biology; accompanied by a very good introduction to the
issues.

444.   Coleman, William. "Morphology between Type
       Concept and Descent Theory." Journal of the
       History of Medicine 31(1976): 149-175.

       Analyzes the retention of portions of type
concept in morphologists' acceptance and definition of
descent theory through the work of Carl Gegenbauer.

445.   Colloque International 'Lamarck'. Tenu au Muséum
       National d'Histoire Naturelle, Paris, les 1-2 et
       3 juillet 1971. Ed. by Joseph Schiller. Paris:
       Blanchard, 1971.

       Contains the papers presented; covers a wide
range of Lamarck's work including one on "L'Échelle des
Etres Chez Lamarck" by Joseph Schiller and one on "Le
Botaniste Lamarck" by M. Guédès.

446.   Farber, Paul Lawrence. "Buffon and Daubenton:
       Divergent Traditions within the Histoire
       Naturelle." Isis 66(1975): 63-74.

       Describes Buffon's and Louis Jean Marie
Daubenton's concepts of natural history and proposes
differences between these concepts as a possible reason
for the end of their collaboration.

447. Farber, Paul Lawrence. "The Transformation of
      Natural History in the Nineteenth Century."
      Journal of the History of Biology 15(1982): 145-
      152.

      Specifies that the changes in natural history
grew out of increased interest and opportunities, an
increased number of naturalists, exponentially and
qualitatively increasing information and collections
data bases, specialization, and new questions (such as
those of species-variety relationship and geographical
distribution).

448. Glick, Thomas F. and David M. Quinlin. "Felix de
      Azara: The Myth of the Isolated Genius in
      Spanish Science." Journal of the History of
      Biology 8(1975): 67-83.

      Discusses characteristics of eighteenth century
Spanish science that perpetuate the myth of the isolated
genius using Felix de Azara and others as examples.
Suggests that Azara was not isolated.

449. Hawkins, Hugh. "Transatlantic Discipleship: Two
      American Biologists and Their German Mentor."
      Isis 71(1980): 197-210.

      Describes the careers of two American students,
John Mason Tyler and Henry Baldwin Ward, who studied in
Germany, primarily at Gothingen under Ernst Heinrich
Ehlers in 1876 and 1888 respectively. Analyzes the
influence on the students of their studies in Germany
and of their mentor Ehlers and the ways in which it
changed throughout their lives.

450. Kessel, Edward L., ed. A Century of Progress in
      the Natural Sciences, 1853-1953. San Francisco:
      California Academy of Sciences, 1955.

      Contains forty essays written for the centennial
of the California Academy of Sciences. Nearly all are
thorough surveys of developments in classification of
plants and animals.

451.   Larson, James L. "Linnaeus and the Natural
        Method." Isis 58(1967): 304-320.

        Explains Linnaeus artificial system of
classification and the reasons for his unsuccessful
attempts with the natural method.

452.   Lenoir, Timothy. "Generational Factors in the
        Origin of 'Romantische Naturphilosophie'."
        Journal of the History of Biology 11(1978): 57-
        100.

        Compares the interpretation of ideal type by the
generation reaching their twenty-fifth birthday from
1790 to 1800 with the previous generation and with the
Naturphilosophen of the same generation, focusing on
Carl Friedrich Kielmeyer, Heinrich Friedrich Link, and
Gottfried Reinhold Treviranus.

453.   Lovejoy, Arthur O. The Great Chain of Being. A
        Study of the History of an Idea. New York:
        Harper, 1960 (reprint) Cambridge, Mass.: 1936.

        Remains a classic; chapters on eighteenth century
biology and on temporalizing the Great Chain of Being
are of particular interest.

454.   Lyon, John, and Phillip R. Sloan, eds. From
        Natural History to the History of Nature:
        Readings from Buffon and His Critics. Notre
        Dame, Ind.: University of Notre Dame Press,
        1981.

        Provides twenty translations selected from
Buffon's writings and from eighteenth century critiques
of them. Contains an introduction to Buffon's work and
an analysis of its role in the transformation of natural
history as well as introductions to the selections.

455.   Lyon, John. "The 'Initial Discourse' to Buffon's
        Histoire Naturelle: The First Complete English
        Translation." Journal of the History of Biology
        9(1976): 133-181.

        Translates Georges Louis LeClerq, Comte de

Buffon's introduction to his Histoire Naturelle.
Analyzes the content of that introduction, which is
titled "Initial Discourse: On the Manner of Studying and
Writing About Natural History."

456.   Mullett, Charles F. "Multum in Parvo: Gilbert
        White of Selborne." Journal of the History of
        Biology 2(1969): 363-389.

        Describes charmingly the natural history of
Gilbert White; cites it as representative of the late
eighteenth century thought and style.

457.   Porter, Charlotte M. "The Concussion of
        Revolution: Publications and Reform at the Early
        Academy of Natural Sciences, Philadelphia, 1812-
        1842." Journal of the History of Biology
        12(1979): 273-292.

        Describes the activities of the Academy's early
field naturalists, such as Alexander Wilson, Thomas
Nuttall, Thomas Say, and Titian Ramsay Peale. Covers
their nationalism, their aversion to foreign artificial
taxonomic systems, their magnificent natural history
publications, and the end of the field naturalists'
leadership of the Academy.

458.   Rehbock, Philip F. The Philosophical Naturalists:
        Themes in Early Nineteenth-Century British
        Biology. Madison: University of Wisconsin
        Press, 1983.

        Analyzes in detail the period between 1830 and
1860 when British philosophical natural historians began
to search for general laws and to discover the laws of
distribution of organisms in space and time. Discusses
in particular the work of Robert Knox, Edward Forbes,
and Hewett Cottrell Watson.

459.   Ritterbush, Phillip C. Overtures to Biology: The
        Speculations of Eighteenth Century Naturalists.
        New Haven: Yale University Press, 1964.

        Discusses two major speculative theories of the
eighteenth century, immanence and botanical analogy.

Describes a trend to empirical and experimental work and the decline of speculation. Examines the work of Erasmus Darwin, Lamarck, Sir Humphrey Davy and John Hunter.

460.   Schofield, Robert E. Mechanism and Materialism: British Natural Philosophy in An Age of Reason. Princeton: Princeton University Press, 1970.

Contains a chapter on "Vital Physiology and Elementary Chemistry."

461.   Urness, Carol, ed. A Naturalist in Russia: Letters from Peter Simon Pallas to Thomas Pennant. Minneapolis: University of Minnesota Press, 1967.

Collects and carefully annotates the eighteenth century correspondence between the German naturalist Peter Simon Pallas and the English naturalist Thomas Pennant.

462.   Von Hagen, Victor Wolfgang. The Green World of the Naturalist: A Treasury of Five Centuries of Natural History in South America. New York: Greenberg, 1948.

Reproduces excerpts from the works of twenty-five natural historians from the sixteenth into the twentieth century who wrote about their adventures with the South American flora and fauna.

NEUROPHYSIOLOGY

463.   Ackerknecht, Erwin H. "The History of the Discovery of the Vegetative (Autonomic) Nervous System." Medical History 18(1974): 1-8.

Identifies the major researchers and their contributions to the understanding of the autonomic nervous system in chronological order from Thomas Willis to H.H. Dale.

464. Andreoli, Armando. Zur Geschichtlichen
      Entwicklung der Neuronen-Theorie. Basel: Benno
      Schwabe, 1961.

      Surveys the history of the nervous system,
focusing on neuronal theories.

465. Brazier, Mary A.B. A History of the Electrical
      Activity of the Brain: the First Half-Century.
      London: Pitman, 1961.

      Outlines briefly the discovery of the link
between electrical activity of the nervous system and
nervous system function and the development of the
concept after Du-Bois-Reymond, through the work of
Richard Caton, Adolf Beck and others.

466. Brazier, Mary A.B. "Rise of Neurophysiology in
      the 19th Century." Journal of Neurophysiology
      20(1957): 212-226.

      Focuses on electrophysiological research
contributions to neurophysiology.

467. Clarke, Edwin and Charles D. O'Malley. The Human
      Brain and Spinal Cord. Berkeley: University of
      California Press, 1968.

      Contains reprints and translations of classic
works on neuroanatomy and neurophysiology accompanied by
biographical sketches and commentary. Covers antiquity
to mid-twentieth century arranged by broad subject areas
such as the neuron, nerve function, the cerebral
convolutions, and brain localization.

468. Clarke, Edwin and K. Dewhurst. An Illustrated
      History of Brain Function. Berkeley: University
      of California Press, 1972.

      Contains chronologically arranged illustrations
of localization of function in the brain, accompanied by
brief explanatory text.

469. Cranefield, Paul F. The Way In and the Way Out.
      Francois Magendie, Charles Bell, and the Roots
      of the Spinal Nerves. New York: Futura, 1975.

        Discusses the priority dispute regarding the
discovery of the sensory and motor roots. Contains
facsimiles of all materials upon which priority claims
could be based, including Charles Bell's extensively
annotated copy of his Ideas of a New Anatomy of the
Brain, two of Bell's articles, one of John Shaw's
articles, and Magendie's articles. Contains excerpts of
other relevant material and an annotated bibliography.

470. French, Richard D. "Some Concepts of Nerve
      Structure and Function in Great Britain, 1875-
      1885: Background to Sir Charles Sherrington and
      the Synapse Concept." Medical History 14(1970):
      154-165.

        Suggests that the invertebrate physiological
research of George G. Romanes and Edward A. Schafer and
Charles Scott Sherrington's exposure to it while a
student of Walter Gaskell at Cambridge influenced
Sherrington's research on the nervous system.

471. Fulton, John Farquhar. Physiology of the Nervous
      System. 3d ed. revised. New York: Oxford
      University Press, c1949, 1951.

        Classic example of a scientific monograph that
provides historical notes preceding each chapter listing
the major steps in the development of the topic, with
citations to the primary works.

472. Garrison, F.H. History of Neurology.
      Springfield: Thomas, 1969.

        Surveys neurology with chapters on neuroanatomy,
neurophysiology and neurochemistry.

473. Haymaker, Webb, and Francis Schiller, comp. and
      ed. Founders of Neurology: One Hundred and
      Forty-Six Biographical Sketches by Eighty-Eight
      Authors. 2d ed. Springfield: Thomas, 1970.

Arranges biographies by research area; includes portraits and references to other biographies.

474. The History and Philosophy of Knowledge of the Brain and Its Functions: An Anglo-American Symposium, London, July 15-17, 1957. Oxford: Blackwell, 1958.

Contains seventeen papers prepared for an historical pre-conference to the First International Congress of Neurological Sciences, Brussels, 1957. Includes seven papers that cover 1700 to date, of which MacDonald Critchley's "On Origins of Language" and Mary A.B. Brazier's "On Electrical Activity of the Nervous System" are of particular interest.

475. Home, Roderick W. "Electricity and the Nervous Fluid." Journal of the History of Biology 3(1970): 235-251.

Discusses Albrecht von Haller's rejection of previous theories on how muscular contraction is effected and his concept of a nervous fluid. Reviews arguments against the electrical fluid and the nervous fluid being the same; explains Luigi Galvani's position that they are the same.

476. Jacyna, L.S. "Principles of General Physiology: The Comparative Dimension to British Neuroscience in the 1830s and 1840s." Studies in History of Biology 7(1984): 47-92.

Analyzes the work in comparative anatomy and physiology from the biological perspective of William B. Carpenter, Richard Grainger, and Samuel Solly. Focuses on their theories, based on the transcendental philosophy of nature, of nervous system structure and function.

477. Keele, K.D. Anatomies of Pain. Oxford: Blackwell, 1957.

Discusses historical concepts of pain and the development of knowledge about its anatomical and physiological basis.

478.   Leys, Ruth.   "Background to the Reflex
        Controversy: William Alison and the Doctrine of
        Sympathy Before Hall."   Studies in History of
        Biology 4(1980): 1-66.

        Examines the conflicting views of the nervous
system by British physiologists in the 1820s just prior
to Marshall Hall's research publications on the anatomy
and physiology of reflexes. Focuses in particular on
the views of William Pulteney Alison, who held a
psychological explanation of reflex actions based on
Robert Whytt's ideas of sympathy.

479.   Liddell, E.G.T.   The Discovery of Reflexes.
        Oxford: Clarendon Press, 1960.

        Outlines chronologically for neurophysiologists
the history of reflexes up to the publication of Charles
Scott Sherrington's The Integrative Action of the
Nervous System. Contains chapters on the "Nerve Cell
and the Microscope," "Animal Electricity," Experimental
Approaches," and "Sherrington and His Times." Includes
a biographical appendix.

480.   Rose, F. Clifford, and W.F. Bynum, ed.   Historical
        Aspects of the Neurosciences. New York: Raven,
        1982.

        Contains forty-six papers presented at the first
meeting of the History of the Neurosciences Group and
published as a festschrift for Macdonald Critchley.

481.   Rothschuh, Karl E., ed.   Von Boerhaave bis Berger:
        Die Entwicklung der kontinentalen Physiologie im
        18. und 19. Jahrhundert mit besonderer
        Berücksictigung der Neurophysiologie.
        Stuttgart: Gustav Fischer, 1964.

        Contains papers presented at a 1962 international
symposium; covers Emil du Bois-Reymond's work on the
electrophysiology of the nerves.

482.   Swazey, Judith P.   "Action Propre and Action
        Commune: The Localization of Cerebral Function."

Journal of the History of Biology 3(1970): 213-234.

Reviews the nineteenth century controversy over action propre and action commune through the phrenology of Franz Gall, the research of Pierre Flourens, and the clinical research of Paul Broca.

483.   Swazey, Judith P. Reflexes and Motor Integration: Sherrington's Concept of Integrative Action. Cambridge, Mass.: Harvard University Press, 1969.

Analyzes Charles Scott Sherrington's development of the concept of integrative action of the nervous system. Describes earlier studies of reflex action; discusses Sherrington's extensive research in its scientific context, and assesses Sherrington's contribution to neurophysiology.

484.   Swazey, Judith P. "Sherrington's Concept of Integrative Action." Journal of the History of Biology 1(1968): 57-89.

Covers the development of the integrative action concept in Charles Scott Sherrington's work from 1884 to 1906 during which he provided an adequate anatomical base and experimentally demonstrated the basis of the integrative action of the nervous system.

485.   Tizard, Barbara. "Theories of Brain Localization from Flourens to Lashley." Medical History 3(1959): 132-145.

Traces research on localization and field theories of brain function from the field theory of Pierre Flourens, through classical localization theories, to Karl Lashley's field theory.

ORNITHOLOGY

486.   Farber, Paul Lawrence. "The Development of Taxidermy and the History of Ornithology." Isis 68(1977): 550-556.

Describes efforts of taxidermists to preserve
ornithological specimens from René Antoine Ferchault de
Réaumur in the mid-eighteenth century to the solution of
the problem at the turn of the century. Makes clear the
importance of having large permanent ornithological
specimen collections as a data base for classification,
bird lists, the study of groups and of geographic
distribution, and as a basis for accurate illustrations
and the designation of type species.

487.   Farber, Paul Lawrence. The Emergence of
       Ornithology as a Scientific Discipline: 1760-
       1850. Dordrecht: Reidel, 1982.

Assesses the study of ornithology in the mid-
eighteenth century and analyzes the development of
ornithology as an independent, scientific zoological
discipline of natural history to mid-nineteenth century.
Covers individuals, institutions, associations,
publications, and research traditions of that period.

488.   Stresemann, Erwin. Ornithology: From Aristotle to
       the Present. Translated by Hans J. and Cathleen
       Epstein. Edited by G. William Cottrell.
       Cambridge: Harvard University Press, 1975.

Translates Stresemann's classic 1951 survey of
the development of the field of ornithology from
description and collection through empiricism and
behavioral studies. Includes a chapter on "Materials
for a History of American Ornithology" by Ernst Mayr.

PALEONTOLOGY

489.   Adams, Frank D. The Birth and Development of the
       Geological Sciences. New York: Dover, 1954.

Reprints of the 1938 classic survey of the
field, with a chapter on the founding of paleontology,
covering in particular the contributions of Georges
Cuvier and William Smith.

490.   Andrews, H.N. The Fossil Hunters: In Search of
       Ancient Plants. Ithaca: Cornell University

Press, 1980.

Reviews in popularized biographical style the
history of paleontologists' efforts to find and explain
dinosaurs.

491. Bourdier, Franck. "Geoffroy Saint-Hilaire Versus
      Cuvier; the Campaign for Paleontological
      Evolution (1825-1838)." In Toward a History of
      Geology, edited by Cecil J. Schneer, 36-61.
      Cambridge: M.I.T. Press, 1969.

Describes the differing positions on evolution
that Cuvier and Geoffroy Saint-Hilaire took from their
student days throughout their careers. Relates Geoffroy
Saint-Hilaire's examination of the "crocodile" of Caen
and his subsequent search for fossil evidence of the
great chain of being.

492. Bowler, Peter J. Fossils and Progress:
      Paleontology and the Idea of Progressive
      Evolution in the Nineteenth Century. New York:
      Science History Publications, 1976.

Analyzes the development of positions on
progressionism as influenced by the growth of
understanding of the fossil record in the nineteenth
century from Georges Cuvier through Charles Darwin.

493. Colbert, Edwin H. Men and Dinosaurs: The Search
      in Field and Laboratory. New York: Dutton,
      1968.

Relates in a biographical, popularized style the
history of paleontologists' worldwide study of
dinosaurs.

494. Conry, Yvette. Correspondence entre Charles
      Darwin et Gaston de Saporta. Précédé d'une
      histoire de la paléobotanique en France en XIXe
      siècle. Paris: Presses Universitaires de
      France, 1972.

Contains the correspondence between Charles

Darwin and Gaston de Saporta; includes a long
introductory essay on the history of paleobotany.

495. Desmond, Adrian J. "Designing the Dinosaur:
      Richard Owen's Response to Robert Edmond Grant."
      Isis 70(1979): 224-234.

Suggests that Richard Owen's creation of the
dinosaur first mentioned in his 1841 speech on Mesozoic
saurians to the British Association was in response to
the perceived threat of Robert Edmond Grant's historical
Lamarckism.

496. Gerstner, Patsy A. "Vertebrate Paleontology, an
      Early Nineteenth Century Transatlantic Science."
      Journal of the History of Biology 3(1970): 137-
      148.

Focuses on the career of Richard Harlan as an
example of post 1830 European respect for the work of
American paleontologists.

497. Rainger, Ronald. "The Continuation of the
      Morphological Tradition: American Paleontology
      1880-1900." Journal of the History of Biology

Provides evidence for the persistence of a
morphological tradition in paleontology. Presents the
work of Othniel Charles Marsh (1832-1899), Edward
Drinker Cope (1840-1897), Alpheus Hyatt (1838-1902), and
students, on embryological and evolutionary changes in
fossils.

498. Rohdendorf, B.B. "The History of
      Paleoentomology." In History of Entomology,
      edited by Ray F. Smith, Thomas E. Mittler, and
      Carroll N. Smith, 155-170. Palo Alto: Annual
      Reviews, 1973.

Surveys the history of paleoentomology;
identifies the issues of completeness, level of
description and comparisons, and age. Specifies which
insect groups, which time periods, and which geographic
areas have been studied and by whom. Identifies the

phylogenetic aspects of the work introduced at the turn of the century.

499. Rudwick, Martin J.S. The Meaning of Fossils.
      Episodes in the History of Paleontology. 2d
      ed. New York: Science History, 1976.

      Interprets paleontology's role in the development
of the concept of biological evolution and attends
carefully to the technical aspects of understanding the
fossil record.

500. Senet, Andre. Man in Search of His Ancestors: The
      Romance of Palaeontology. London: Allen &
      Unwin; New York: McGraw-Hill, 1956.

      Remains a classic popularized survey of the
history of paleontology.

501. Siesser, W.G. "Christian Gottfried Ehrenberg:
      Founder of Micropaleontology." Centaurus
      25(1981): 166-188.

      Outlines Christian Gottfried Ehrenberg's
contributions to the microscopic study of diatoms,
radiolarians, foraminifera, and other microfossils.
Notes Ehrenberg's identification of the role that
microfossils play in rock-building.

502. Todes, Daniel P. "V. O. Kovalevskii: The Genesis,
      Content, and Reception of His Paleontological
      Work." Studies in History of Biology 2(1978):
      99-165.

      Discusses the career and the contributions to
evolutionary paleontology of the nineteenth century
Russian paleontologist Vladimir Onufrievich Kovalevski.

503. Zittel, Karl von. History of Geology and
      Palaeontology. Weinheim: J. Cramer, 1962.

      Contains a chapter on the history of paleontology
identifying major individuals and their

contributions, associations, trends and research areas.
Reprint of what remains an old classic in the field.

PHYSICAL ANTHROPOLOGY

504.   Fleagle, John G., and William L. Jungers.   "Fifty
        Years of Higher Primate Phylogeny." In A
        History of American Physical Anthropology, 1930-
        1980, edited by Frank Spencer, 187-230. New
        York: Academic Press, 1982.

        Covers the history of primate phylogeny from 1930
and the first meeting of the American Association of
Physical Anthropology to 1980. Discusses both the
contributions of important individuals, such as W.K.
Gregory and Henry Fairfield, and the theoretical and
methodological issues.

505.   Hulse, F.S. "Habits, Habitats, and Heredity: A
        Brief History of Studies in Human Plasticity."
        American Journal of Physical Anthropology
        56(1981): 495-501.

        Recounts the nature/nurture controversy in
physical anthropology and describes research on the
topic. Mentions especially changes in bodily form.

506.   Slotkin, James Sydney, ed. Readings in Early
        Anthropology. Chicago: Aldine, 1965.

        Compiles excerpts from classic anthropological
works beginning with the twelfth century and providing
four chapters on the eighteenth century. Covers
physical anthropology.

507.   Spencer, Frank, ed. A History of American
        Physical Anthropology, 1930-1980. New York:
        Academic Press, 1982.

        Surveys the recent history of physical
anthropology; covers theoretical and methodological
research issues in such areas as primate neuroanatomy,
molecular anthropology and primate field studies, and

includes a chapter on the effects of funding patterns on the discipline.

508. Washburn, S.L. "One Hundred Years of Biological Anthropology." In One Hundred Years of Anthropology, edited by J.O. Brew, 97-115. Cambridge: Harvard University Press, 1968.

Reviews the nature of evolutionary issues for anthropology in a paper presented for the centennial of the Peabody Museum.

PHYSIOLOGY

509. Albury, William Randall. "Experiment and Explanation in the Physiology of Bichat and Magendie." Studies in the History of Biology 1(1977): 47-131.

Compares and contrasts the research methodologies and the explanatory systems of Xavier Bichat and François Magendie.

510. Benzinger, Theodor H., ed. Temperature: Part I, Arts and Concepts; Part II, Thermal Homeostasis. Stroudsburg, Pennsylvania: Dowden, Hutchinson & Ross, 1977.

Contains reprints and translations of forty-one classic papers on the thermal homeostasis of man through mid-twentieth century, including contributions to instrumentation. Provides introductions and brief biographical information for the thirteen sections, which include calorimetry, thermoregulatory responses, thermometry, firing central thermoreceptors, and neurochemical mechanisms.

511. Blasius, Wilhelm, John W. Boylan, and Kurt Kramer, eds. Founders of Experimental Physiology. Munich: Lehmanns, 1971.

Contains facsimiles of significant contributions to physiology by eleven physiologists including Luigi Galvani, Julius Robert von Mayer, Hermann von Helmholtz,

Carl Ludwig, Adolf Fick, Claude Bernard, Otto Frank, and
Julius Bernstein.

512.   Brooks, Chandler McC., and Paul F. Cranefield,
       eds. The Historical Development of
       Physiological Thought: a Symposium Held at the
       State University of New York Downstate Medical
       Center. New York: Hafner, 1959.

Includes chapters on basic physiological
concepts, with an introduction by Owsei Temkin on "The
Dependence of Medicine upon Basic Scientific Thought."
Treats four broad areas: "The Basis of Integrative
Function and Human Behavior," "Humoral Transport and
Integrative Function," "Mechanistic Thought, Energetics
and Control in Biology," and "The Vital Process and the
Disease State," with subsections contributed by
physiologists.

513.   Brooks, Chandler McC., Kiyomi Koizumi, and James
       O. Pinkston, eds. The Life and Contributions of
       Walter Bradford Cannon, 1871-1945. New York:
       State University of New York Press, 1975.

Contains ten papers presented at a centennial
symposium on Walter Bradford Cannon's influence on the
development of twentieth century physiology. Covers the
physiology of the digestive system, of the autonomic
nervous system, of the expression of emotions, and of
the maintenance of homeostasis and reactions to stress,
with the papers including varying amounts of historical
background.

514.   Brown, Theodore M. "From Mechanism to Vitalism in
       Eighteenth-Century English Physiology." Journal
       of the History of Biology 7(1974): 179-216.

Discusses the role that changes in medical theory
and the medical profession played in the change from
mechanism to vitalism in physiology.

515.   Coleman, William. "Bergmann's Rule: Animal Heat
       as a Biological Phenomenon." Studies in History
       of Biology 3(1979): 67-88.

Examines Carl Bergmann's 1840s inductive work on the production and regulation of animal heat that resulted in 1847 in the formulation of his rule on size distribution among members of closely related groups of warm-blooded animals. Demonstrates that Bergmann established the conceptual framework upon which later research was carried out.

516. Cross, Stephan. "John Hunter, the Animal Oeconomy, and Late Eighteenth Century Physiological Discourse." Studies in History of Biology 5(1981): 1-110.

Analyzes the texts and collection of John Hunter to reveal Hunter's conceptual framework and its relationship to late eighteenth century biology. Categorizes Hunter's physiological work as pre-biological.

517. Delaporte, Francois. "Theories of Osteogenesis in the Eighteenth Century." Journal of the History of Biology 16(1983): 343-360.

Attributes research and contributions on the role of the periosteum in bone formation and growth to Henri-Louis Duhamel du Monceau, John Hunter, Michele Troja, and Georges Cuvier.

518. Duchesneau, François. La Physiologie des Lumières: Empirisme, Modèles et Théories. The Hague: Martinus Nijhoff, 1982.

Analyzes in depth the physiological theories of the enlightenment.

519. Fenn, Wallace O., ed. History of the International Congresses of Physiological Sciences, 1889-1968. Baltimore: Waverly, 1968.

Contains a reprint of K.J. Franklin's "A Short History of the International Congresses of Physiologists 1889-1938" from Annals of Science 3(1938): 244-335, which describes the congresses and has thirty-five photographs of notables and attendees. Includes Yngve Zotterman's recounting of "The Minnekahda Voyage" in

1929 when European physiologists chartered a ship to
attend the congress in America, and short descriptions
of the congresses and their arrangement from 1938-1968,
written by Presidents or Local Chairs. Ends with a
chapter on the International Union of Physiological
Sciences assuming responsibility for the congresses in
1953.

520.   Fishman, Alfred P., and Dickinson W. Richards,
       eds. Circulation of the Blood: Men and Ideas.
       New York: Oxford University Press, 1964.

       Contains twelve chapters on cardiovascular
physiology written by eminent men in the field for an
audience of physiologists and historians. Surveys major
contributions in such areas as heart muscle, vasomotor
control and the regulation of blood pressure, or
physiological changes in the circulation after birth.

521.   French, Richard D. "Darwin and the Physiologists,
       or the Medusa and Modern Cardiology." Journal
       of the History of Biology 3(1970): 253-274.

       Demonstrates the influence of Darwin and
Darwinian theory on late nineteenth century physiology
as exemplified in the research of George Romanes, Walter
Gaskell, Michael Foster, and John Burdon Sanderson on
the evolutionary relationship between structure and
function in nervous units.

522.   Fulton, John Farquhar. Physiology. New York:
       Hoeber, 1931.

       Remains a standard early survey of circulation,
respiration, and digestion. Contains a chapter on the
beginnings of teaching laboratories in the nineteenth
century.

523.   Fulton, John Farquhar, and Leonard G. Wilson, eds.
       Selected Readings in the History of Physiology.
       2d ed. Chicago: C.C. Thomas, 1966.

       Selects and carefully annotates and illustrates
classic passages from twelve areas of physiology -
biophysical forerunners, vascular system: discovery of

the circulation, blood capillaries, respiration: neurophysiology and chemistry, digestion, muscle and peripheral nerves, central nervous system, homeostasis, kidney: concepts of renal function, sexual generation, endocrinology, and vitamins.

524. Geison, Gerald L. Michael Foster and the Cambridge School of Physiology: The Scientific Enterprise in Late Victorian Society. Princeton: Princeton University Press, 1978.

Analyzes the state of English physiology from 1840 to 1870, particularly at Cambridge. Discusses Michael Foster and demonstrates the importance of Foster's own research and his leadership in establishing a center for physiological research at Cambridge.

525. Goodfield, G. June. The Growth of Scientific Physiology: Physiological Method and the Mechanist-Vitalist Controversy, Illustrated by the Problems of Respiration and Animal Heat. London: Hutchinson, 1960.

Discusses the development of concepts of thermoregulation as an exemplar of the development of scientific physiology, including an experimental base, and of the philosophical positions of the mechanists and vitalists.

526. Goodfield-Toulmin, June. "Some Aspects of English Physiology: 1780-1840." Journal of the History of Biology 2(1969): 283-320.

Notes the importance of interrelationships between physiology and philosophy set in social and political contexts. Discusses in particular the explanation of the phenomena of life based on the theory of the vital principal which was an issue in physiology and medicine from 1780-1830 in England. Comments on the use of analogy to Newtonian gravitation to justify it. Details the case of William Lawrence.

527. Grande, Francisco and Maurice B. Visscher, eds. Claude Bernard and Experimental Medicine. Cambridge, Mass.: Schenkman, 1967.

Contains fourteen papers presented at 1965 symposium commemorating the centenary of the publication of Claude Bernard's An Introduction to the Study of Experimental Medicine and the first English translation of his Cahier Rouge. Includes "Historical Phases in the Influence of Bernard's Scientific Generalizations in England and America" By E. Harris Olmsted, "Origins of the Concept of Milieu Interieur" by Frederick L. Holmes, and an English translation of Bernard's Cahier Rouge.

528. Hall, Thomas Steele. Ideas of Life and Matter. Studies in the History of General Physiology 600 B.C.-1900 A.D. 2 vols. Chicago: University of Chicago Press, 1969.

Includes in volume two "From the Enlightenment to the End of the Nineteenth Century" chapters on "Mechanism and Vitalism," on "Tissue, Cell, and Molecule," and on "The Physical Basis of Life."

529. Heim, Roger, ed. Les Concepts de Claude Bernard sur le Milieu Interior. Paris: Masson, 1967.

Contains the proceedings of an international symposium held on the anniversary of the publication of Claude Bernard's L'Introduction à l'Etude de la Medecine Experimentale.

530. Hodgkin, Alan L., Andrew F. Huxley, W. Feldberg, W.A.H. Rushton, R.A. Gregory, and R.A. McChance, eds. The Pursuit of Nature: Informal Essays on the History of Physiology. Cambridge: Cambridge University Press, 1977.

Contains six essays on various aspects of physiological research written by participants for the Centenary of the Physiological Society. Covers mainly twentieth century research on electrophysiology, muscle, synaptic and neuromuscular transmission by acetylcholine, vision, gastrointestinal hormones, and perinatal physiology.

531. Holmes, Frederick L. Claude Bernard and Animal Chemistry. The Emergence of a Scientist. Cambridge: Harvard University Press, 1974.

Examines Claude Bernard's research on digestion and nutrition during the period 1842-1848. Discusses his scientific development and the scientific context in which he worked.

532.   Holmes, Frederick L. "Joseph Barcroft and the Fixity of the Internal Environment." Journal of the History of Biology 2(1969): 89-122.

Discusses the ambiguity of the implicit influence of the idea of "milieu interior" on the research of the British physiologist Joseph Barcroft during the early twentieth century.

533.   Kepner, G.R., ed. Cell Membrane Permeability and Transport. Stroudsburg, Pa.: Dowden, Hutchinson & Ross, 1979.

Includes classic nineteenth and twentieth century (up to 1960) articles on water and nonelectrolyte permeability and transport and on sodium and potassium permeability and transport, accompanied by editorial comments on the two sections.

534.   Koshtoyants, Kh.S. Essays on the History of Physiology in Russia. Edited by Donald B. Lindsley. Translated by David P. Boder, Kristen Hanes and Natalie O'Brian. Washington, D.C.: American Institute of Biological Science, 1964.

Covers Russian physiology from the eighteenth century into the twentieth. Emphasizes the research of I.M. Sechenov, then Ivan Petrovich Pavlov, but also covering the research of many other contributors.

535.   Langley, L.L., ed. Homeostasis, Origins of the Concept. Stroudsburg, Pennsylvania: Dowden, Hutchinson & Ross, 1973.

Contains facsimiles and translations of twenty-two classic contributions to the understanding of homeostasis. Includes brief introductory remarks.

536. Lipman, Timothy O. "The Response to Liebig's
      Vitalism." Bulletin of the History of Medicine
      40(1966): 511-524.

      Notes an initial lack of response to the use of
vital force in Justus von Liebig's book, Animal
Chemistry. Examines the later critical response and
discusses the elimination of vital force as a valid
concept through the work of Virchow, Bernard, and
Moleschott among others.

537. Lipman, Timothy O. "Vitalism and Reductionism in
      Liebig's Physiological Thought." Isis 58(1967):
      167-185.

      Examines vitalism and reductionism in Justus von
Liebig's organic chemistry, views on fermentation, and
physiology. Explains his application of vitalism to
physiology.

538. Lomax, Elizabeth. "Historical Development of
      Concepts of Thermoregulation." In Body
      Temperature: Regulation, Drug Effects and
      Therapeutic Implications, edited by Peter Lomax
      and Edward Schonbaum, 1-24. New York: Marcel
      Dekker, 1979.

      Outlines briefly contributions to the development
of concepts of thermoregulation from the ancients
through the early twentieth century.

539. Mazumdar, Pauline H.M. "Johannes Müller on the
      Blood, the Lymph, and the Chyle." Isis
      66(1975): 242-253.

      Describes Johannes Müller's research on fibrin
and red blood corpuscles as significant for its
challenge to Naturphilosophisch and for its introduction
of emphasis on experiment in Germany.

540. Mendelsohn, Everett. Heat and Life: The
      Development of the Theory of Animal Heat.
      Cambridge: Harvard University Press, 1964.

      Discusses the critical role of physical and

mechanical models in the development of biological
concepts and theories about animal heat.

541. Parascandola, John. "Organismic and Holistic
      Concepts in the Thought of L.J. Henderson."
      Journal of the History of Biology 4(1971): 63-
      113.

      Discusses Lawrence Joseph Henderson's ideas on
the organization, regulation, and equilibrium of
systems, and their development throughout his education,
career, and research in a number of fields.

542. Pickstone, John V. "Absorption and Osmosis:
      French Physiology and Physics in the Early
      Nineteenth Century." Physiologist 20(1977): 30-
      37.

      Compares and contrasts the work of François
Magendie, Michel Fodera, and Henri Dutrochet on the
problem of absorption.

543. Rothschuh, Karl E. History of Physiology.
      Translated and edited by Guenter B. Risse. New
      York: Robert E. Krieger, 1973.

      Surveys physiologists and physiology from
antiquity into the twentieth century. Includes a
chapter on "Physiology of the Enlightenment," on
nineteenth century physiology, on "Johannes Müller, Carl
Ludwig, and their Circle of Students," and on
"Nineteenth and Twentieth Century Physiology in Western
Europe, America, and Russia."

544. Schiller, Joseph and Tetty Schiller. Henri
      Dutrochet: Le Materialisme Mécaniste et la
      Physiologie Générale. Paris: Albert Blanchard,
      1975.

      Presents previously unpublished letters and
autobiographical sketches (Notice sur Ma Vie and Notice
sur Mes Ouvrages) accompanied by the editor's
biographical notes and discussions of Dutrochet's work.

545. Schultheisz, E., ed. History of Physiology.
     Oxford: Pergamon, 1981.

     Contains ten brief historical papers presented at
the twenty-eighth International Congress of
Physiological Sciences, including ones on comparative
physiology, psychophysiology, the anatomy and physiology
of the pig fetus and placenta, and on Jan Nepomuk
Czermak.

546. Sharpey-Schafer, Edward. History of the
     Physiological Society During Its First Fifty
     Years, 1876-1926. London: Cambridge University
     Press, 1927.

     Outlines the founding of the Society; includes
biographical information about its founding members and
a listing of the activities of its meetings.

547. Sherman, Paul D. Colour Vision in the Nineteenth
     Century: The Young-Helmholtz-Maxwell Theory.
     Bristol: Hilger, 1981.

     Covers in detail theories and research on color
vision and color perception in the period 1800-1860.
Focuses on the work of Thomas Young, Hermann von
Helmholtz, and James Clerk Maxwell on the physiology of
color vision, additive and subtractive color mixing, and
experimentally measured subjective color sensations.

POPULATION GENETICS

548. Norton, Bernard J. "Metaphysics and Population
     Genetics: Karl Pearson and the Background to
     Fisher's Multi-factorial Theory of Inheritance."
     Annals of Science 32(1975): 537-553.

     Suggests that Karl Pearson's reason for lack of
interest in a Mendelian paradigm was metaphysical.

549. Provine, William B. "Origins of The Genetics of
     Natural Population Series." In Dobzhansky's
     Genetics of Natural Populations. I-XLIII,
     edited by R.C. Lewontin, John A. Moore, William

B. Provine, and Bruce Wallace; 1-92. New York:
Columbia University Press, 1981.

Explains the role of Theodosius Dobzhansky, A.H.
Sturtevant, and Sewall Wright in the initiation of the
Genetics of Natural Populations series.

550. Provine, William B. The Origins of Theoretical
     Population Genetics. Chicago: Chicago
     University Press, 1971.

Traces the development of population genetics
from Darwin to Ronald Alymer Fisher, Sewall Wright, and
J.B.S. Haldane, whose work effected a synthesis.
Reviews the controversies and theories of continuous or
discontinuous evolution.

551. Provine, William B. "The Role of Mathematical
     Population Geneticists in the Evolutionary
     Synthesis of the 1930s and the 1940s. Studies
     in the History of Biology 2(1978): 167-192.

Evaluates the role of mathematical population
genetics in the evolutionary synthesis of 1930-1950.
Describes briefly the models of Sergei Chetverikov, R.A.
Fisher, J.B.S. Haldane, and Sewall Wright and discusses
four ways in which they had considerable influence.

552. Roger, Jacques, ed. R.A. Fisher et l'Histoire de
     la Genetique des Populations. Paris: Editions
     Albin Michel, 1981.

Contains essays written to celebrate the
publication of R.A. Fisher's book The Genetical Theory
of Natural Selection; covers theories, research, and
individuals significant in the history of population
genetics.

553. Spiess, Eliot B. Papers on Animal Population
     Genetics. Boston: Little, Brown, 1962.

Reprints thirty-seven classic articles on animal
experimental population genetics, with primarily
evolutionary orientations, from the 1940s through 1961.

554. Wright, Samuel. "The Foundations of Population
      Genetics." In Heritage from Mendel, edited by
      R. Alexander Brink, 245-263. Madison:
      University of Wisconsin Press, 1967.

      Outlines early work briefly and then reports
research from 1911 to 1940.

PSYCHOLOGY

555. Boakes, Robert. From Darwin to Behaviorism:
      Psychology and the Minds of Animals. Cambridge:
      Cambridge University Press, 1984.

      Analyzes the history of research on animal
behavior from 1870 to 1930 and its relationship to
psychology. Beginning with evolutionary behavior at the
time of Darwin the work covers such topics as instinct,
reflex action and the nervous system, conditioned
reflexes, comparative psychology, and the nature/nurture
problem.

556. Boring, Edwin G. A History of Experimental
      Psychology. 2d ed. New York: Appleton, 1950.

      Focuses on psychology from 1860 to 1910 and its
emergence as an experimental science from a largely
biographical view.

557. Burnham, John C. "The Mind-Body Problem in the
      Early Twentieth Century." Perspectives in
      Biology and Medicine 20(1977): 271-284.

      Discusses changes in ideas about the mind-body
problem in the period from 1890 to 1930, during which
the main focus was on the body and included developments
in the understanding of the autonomic nervous system and
the endocrines.

558. Daston, Lorraine J. "British Responses to Psycho-
      Physiology, 1860-1900." Isis 69(1978): 192-208.

      Discusses the issues, including subject matter
and methodology, involved in establishing psychology in

Britain as an independent discipline based on the
natural sciences rather than the philosophical.

559. Davis, Audrey B., and Uta C. Merzbach. Early
     Auditory Studies: Activities in the Psychology
     Laboratories of American Universities.
     Washington, D.C.: Smithsonian Institution,
     1975.

     Describes the development of American research on
audition by psychologists from 1875 to 1930.

560. Dennis, Wayne. Readings in the History of
     Psychology. New York: Appleton, 1955.

     Reprints sixty-one classic works, or excerpts
from them, in chronological order by date of
publication. Emphasis is on physiological and
experimental works by authors such as Pierre Flourens,
Thomas Young, Hermann von Helmholtz, Gustav Theodor
Fechner, and Ivan Petrovitch Pavlov.

561. Esper, Erwin A. "Psychology as a Biological
     Science." In A History of Psychology, 281-340.
     Philadelphia: Saunders, 1964.

     Discusses the historical relationships between
psychology and physiology.

562. Fearing, Franklin. Reflex Action. A Study in the
     History of Physiological Psychology. Baltimore:
     Williams & Wilkins, 1930.

     Surveys chronologically the physiological and
psychological theories and research behind the
development of the concept of reflex action.

563. Gray, Philip Howard. "Spalding and His Influence
     on Developmental Behaviour." Journal of the
     History of the Behavioral Sciences 3(1967): 168-
     179.

     Describes Douglas Alexander Spalding's nineteenth
century animal behavior research, especially

that on visual and auditory development and on the
maturation of behavioral function.

564.   Grinder, Robert E. A History of Genetic
        Psychology: The First Science of Human
        Development. New York: Wiley, 1967.

Covers the concepts of heredity and growth from
Aristotle through G. Stanley Hall and his program at
Clark University. Suggests the latter was the beginning
of the science of human development.

565.   Gruber, Howard E. Darwin on Man: A Psychological
        Study of Scientific Creativity. Together with
        Darwin's Early and Unpublished Notebooks,
        transcribed and annotated by Paul H. Barrett.
        London: Wildwood House, 1974.

Reconstructs the development of Darwin's thought.
Includes his previously unpublished M and N notebooks,
miscellaneous notes, essay on theology and natural
selection, and questions for Mr. Wynne.

566.   Haraway, Donna Jeanne. "Signs of Dominance: From
        a Physiology to a Cybernetics of Primate
        Society, C.R. Carpenter, 1930-1970." Studies in
        History of Biology 6(1983): 129-219.

Details Clarence Ray Carpenter's contributions to
the biological and social theory of dominance. Analyzes
the development of Carpenter's work, which was rooted in
psychology and physiology and used laboratory and field
research methods. Discusses aspects of his work, such
as the use of semiotics and sociometry.

567.   Hearnshaw, L.S. A Short History of British
        Psychology 1840-1940. London: Methuen, 1964.

Outlines British psychology from 1840 to 1940 at
an introductory level. Includes chapters on
physiological, neurophysiological, evolutionary,
psychometric, and comparative psychology.

568. Herrnstein, Richard J. and Edwin G. Boring, eds.
      A Sourcebook in the History of Psychology.
      Cambridge: Harvard University Press, 1965.

      Contains selected classic readings, in
translation where necessary, in fifteen areas of
psychology from antiquity to 1900. Provides an
introduction to each excerpt and concentrates on
experimental and quantitative psychology.

569. Hilgard, Ernest R. "Psychology After Darwin."   In
      Evolution After Darwin, edited by Sol Tax, v.
      II, 269-287. Chicago: University of Chicago
      Press, 1960.

      Discusses Darwin's influence on research in
comparative psychology (evolutionary animal behavior),
emotional expression, and individual differences.

570. Jacyna, L.S. "The Physiology of Mind, the Unity
      of Nature, and the Moral Order in Victorian
      Thought." British Journal for the History of
      Science 14(1981): 102-132.

      Identifies three paths to a biological
psychology: the physiological, zoological, and physical.
Focuses on the differing views of Thomas Laycock and
W.B. Carpenter.

571. Kantor, J.R. The Scientific Evolution of
      Psychology. 2 vols. Chicago, Ill. and
      Granville, Ohio: Principia Press, 1963.

      Traces the development of psychology into a
science. Volume two contains chapters on experimental,
physiological, evolutionary, and behavioristic
psychology.

572. Klein, D.B. A History of Scientific Psychology:
      Its Origins and Philosophical Backgrounds. New
      York: Basic Books, 1970.

      Reviews the history of psychology from the Greeks
through the early twentieth century. Focuses in the
final three chapters on scientific psychology with

emphasis on Johann Friedrich Herbart, Rudolf Hermann
Lotze, Alexander Bain, and Wilhelm Wundt.

573. Lindzey, Gardner, ed. A History of Psychology in
     Autobiography. 6 vols. Englewood Cliffs:
     Prentice-Hall, 1974.

Contains historical and autobiographical
interviews with fifty-eight significant psychologists;
an ongoing project.

574. McGuigan, F.J. "The Historical Development of
     Cognitive Psychophysiology: Theory and
     Measurement." In History of Physiology, edited
     by E. Schultheisz. Oxford: Pergamon, 1981.

Describes psychophysiological theories from
ancient times, but emphasizing nineteenth and twentieth
century theories and twentieth century efforts to
measure psychophysiological events.

575. Mecacci, Luciano. Brain and History: The
     Relationship Between Neurophysiology and
     Psychology in Soviet Research. Translated by
     Henry A. Buchtel. New York: Brunner/Mazel,
     1979.

Surveys nineteenth and twentieth century Russian
research on the physiological basis of psychological
processes, including the work of Ivan M. Sechenov,
Vladimir M. Bekhterev, Ivan P. Pavlov, Konstantin M.
Bykov, Aleksandr R. Luria.

576. Misiak, H.M., and Virginia S. Sexton. History of
     Psychology: An Overview. New York: Grune &
     Stratton, 1966.

Surveys psychology at an introductory level
focusing on mid-nineteenth century and after and
emphasizing scientific aspects such as methodology and
physiological roots.

577. Pastore, N. Selective History of Theories of
     Visual Perception: 1650-1950. New York: Oxford

University Press, 1971.

Discusses empiristic theories of visual perception, focusing on their answers to the question of how we see objects as we do.

578. Pauly, Philip J. "The Loeb-Jennings Debate and the Science of Animal Behavior." Journal of the History of the Behavioral Sciences 17(1981): 504-515.

Discusses the differences between Jacques Loeb and Herbert Spencer Jenning's views on the nature and purpose of the biological study of behavior as exemplified in their controversy over research on invertebrate tropisms.

579. Porges, Stephen W., and Michael G.H. Coles, ed. Psychophysiology. Stroudsburg, Pennsylvania: Dowden, Hutchinson, and Ross, 1976.

Reprints, in translation where necessary, twenty-four classic papers on psychophysiology. Arranges these in four areas: "Methodology: Measures and Measurement," "Arousal Theory," "Orienting Reflex and Attention," and "Emotion and Autonomic Conditioning." Includes an introduction and comments on papers.

580. Rieber, R.W., and Kurt Salzinger, eds. Psychology: Theoretical-Historical Perspectives. New York: Academic Press, 1980.

Discusses the history of psychology in thirteen chapters written by psychologists. Begins with European and nineteenth century influences on the development of psychology as a science, with a number of chapters emphasizing the empirical and experimental nature of psychology. Includes a chapter on Darwin's psychology and its relation to evolution.

581. Sahakian, William S., ed. History of Psychology. Itasca, Ill.: Peacock, 1968.

Contains selected classic readings, in translation where necessary, arranged chronologically, then by nationality and specialty, then by significant individuals. Includes selections from experimental, physiological, and evolutionary psychology.

582. Smith, C.U.M. "Charles Darwin, the Origin of Consciousness, and Panpsychism." Journal of the History of Biology 11(1978): 245-267.

Identifies Charles Darwin's thoughts on consciousness in his writings and analyzes the development of his position.

583. Stephens, Lester D. "Joseph Le Conte and the Development of the Physiology and Psychology of Vision in the United States." Annals of Science 37(1980): 303-321.

Discusses the contributions of Joseph Le Conte to the psychological and physiological study of human vision. Focuses on his book, Sight, his interactions with psychologists such as James McKeen Cattell, and the problems Le Conte's Lamarckian views caused with his visual theories.

584. Wasserman, Gerald S. Color Vision: An Historical Introduction. New York: Wiley, 1978.

Presents the fundamental concepts of color vision from Aristotle to mid-twentieth century.

585. Woodward, William R. "From Association to Gestalt: The Fate of Hermann Lotze's Theory of Spatial Perception, 1846-1920." Isis 69(1978): 572-582.

Describes the development of Hermann Lotze's theory of spatial perception and defines and describes a continuum of responses to it from radical empirism to nativism. Comments on the role of theory in establishing research problems for psychology.

586. Young, Robert Maxwell. Mind, Brain and Adaptation in the Nineteenth Century. Cerebral Localization and Its Biological Context from

<u>Gall to Ferrier</u>. Oxford: Clarendon Press, 1970.

Draws together the strands of nineteenth century
contributions to psychological issues, including the
beginnings of attempts to identify and localize the
functions of the brain by Franz Joseph Gall, the
introduction of the experimental method, of evolution,
of association psychology, and of clinical neurology, to
the work of David Ferrier on the identification and
localization of brain functions.

587. Young, Robert Maxwell. "Scholarship and the
      History of the Behavioural Sciences." <u>History</u>
      <u>of Science</u> 5(1966): 1-51.

Reviews thoroughly and thoughtfully the growing
body of literature on the history of the behavioral
sciences; suggests research areas, and comments on
historiography.

RESPIRATORY SYSTEM

588. Comroe, Julius H. ed. <u>Pulmonary and Respiratory</u>
      <u>Physiology</u>. 2 vols. Stroudsburg, Pennsylvania:
      Dowden, Hutchinson, and Ross, 1976.

Contains facsimiles of forty-four and portions of
seventy-two classic articles arranged in six major
categories: "Air, Oxygen, and Carbon Dioxide;"
"Mechanical Properties of the Lungs and Thorax;" "The
Pulmonary Circulation;" "Regulation of Breathing;"
"Pulmonary Function in Man;" and "Oxygen and Carbon
Dioxide: Measurement and Transport." Comments by the
editor introduce each category.

589. Culotta, Charles A. "Tissue Oxidation and
      Theoretical Physiology: Bernard, Ludwig and
      Pfluger." <u>Bulletin of the History of Medicine</u>
      44(1970): 109-140.

Explains clearly the research and theoretical
contributions of Claude Bernard, Carl Ludwig, and Eduard
Pfluger to the understanding of tissue oxidation.

590.   Gottlieb, Leon S. A History of Respiration.
        Springfield: Thomas, 1964.

        Outlines briefly on a popular level a few of the
major events in the history of respiration.

591.   Hall, Diana Long. "The Iatromechanical Background
        of Lagrange's Theory of Animal Heat." Journal
        of the History of Biology 4(1971): 245-248.

        Traces the roots of Italian mathematician Joseph
Louis de Lagrange's interest in the physiology of
respiration to his early years in Turin where he joined
Giovanni Francesco Cigna and the Comte de Saluces in
research and in founding the Turin Philosophical Society
and its journal.

592.   Hall, Diana Long. Why Do Animals Breathe?
        Physiological Problems and Iatromechanical
        Research in the Early Eighteenth Century. New
        York: Arno, 1981.

        Reviews in detail research efforts to understand
the physiology of respiration; emphasizes the work of
John Mayow, Hermann Boerhaave, and Albrecht von Haller.

593.   Keilin, David. The History of Cell Respiration
        and Cytochrome. Cambridge: Cambridge University
        Press, 1966.

        Outlines the history of respiration through 1844,
with a chapter on Charles Alexander MacMunn and
myohaematin and histohaematin. Discusses research on
cytochrome, including the author's own, in the final
seven chapters.

                    TAXONOMY

594.   Anderson, Lorin. "Charles Bonnet's Taxonomy and
        Chain of Being." Journal of the History of
        Ideas 37(1976): 45-58.

        Describes and explains Charles Bonnet's taxonomy
and chain of being in the context of eighteenth century

natural history. Suggests areas where his ideas are
shifting towards the coming biology.

595. Appel, Toby Anita. "Henri de Blainville and the
     Animal Series: A Nineteenth-Century Chain of
     Being." Journal of the History of Biology
     13(1980): 291-319.

     Describes Henri de Blainville's Animal Series and
its formulation and modification. Discusses the roles
of Georges Cuvier, Jean-Baptiste Lamarck, and Geoffroy
Saint-Hilaire in the development of Blainville's ideas
and on his career. Covers the roles of philosophical
anatomy and religion in the formulation of Blainville's
concepts.

596. Broberg, Gunnar. "Homo Sapiens: Linnaeus's
     Classification of Man." In Linnaeus, The Man
     and His Work, edited by Tore Frangsmyr, 156-194.
     Berkeley: University of California Press, 1983.

     Discusses Linnaeus' classification of man with
the animals in his zoological system, his proposal of
additional human species between man and apes, and the
reaction of his contemporaries.

597. Di Gregorio, Mario A. "In Search of the Natural
     System: Problems of Zoological Classification in
     Victorian Britain." History and Philosophy of
     the Life Sciences 4(1982): 225-254.

     Discusses Linnaeus' artificial system and the
natural systems of classification following it.

598. Farber, Paul Lawrence. "The Type-Concept in
     Zoology during the First Half of the Nineteenth
     Century." Journal of the History of Biology
     9(1976): 93-119.

     Demonstrates the significance and complexity of
the type concept by explaining three: classification,
collection, and morphological type concepts, and by
discussing various uses and interpretations.

599. Mayr, Ernst. "Illiger and the Biological Species
      Concept." Journal of the History of Biology
      1(1968): 163-178.

      Summarizes the issues in defining "biological
species." Identifies Johann Carl Wilhelm Illiger as the
sole turn of the century naturalist who wrote at length
on the species problem. Provides a translation of
Illiger's essay "Thoughts on the Concepts Species and
Genus in Natural History."

600. Rudolph, Emanuel D. "The Introduction of the
      Natural System of Classification of Plants to
      Nineteenth Century American Students." Archives
      of Natural History 10(1982): 461-468.

      Outlines the teaching of Linneus' artificial
system of botanical classification by such American
naturalists as Benjamin Smith Barton and Amos Eaton and
its replacement in textbooks and classrooms by the
natural system taught by John Torrey and Asa Gray.

601. Schiller, Joseph. Physiology and Classification:
      Historical Relations. Paris: Maloine, 1980.

      Discusses attempts to classify animals by
structure and attempts to understand the functions of
animals and how these two aims have interacted from
Aristotle to Claude Bernard, with emphasis on the
nineteenth century.

602. Sloan, Phillip R. "The Buffon-Linnaeus
      Controversy." Isis 67 (1976): 356-375.

      Reinterprets Buffon's criticism of Linnean
taxonomy based on an examination of Buffon's
philosophical position, basic taxonomic concepts, and
methodological position.

603. Smit, Piet, and R.J.Ch.V. ter Laage, eds. Essays
      in Biohistory. Utrecht: International
      Association for Plant Taxonomy, 1970.

      Contains twenty historical papers written for the
sixtieth birthday of Frans Verdoorn. Covers a wide

range of time periods and topics although most are
botanical, including a history of taxonomic studies in
Hevea, tributes to Julius Sachs, Sebastien Vaillant,
August de Saint-Hilaire, and articles on botanical
illustration.

604. Winsor, Mary Pickard. "Barnacle Larvae in the
      Nineteenth Century: A Case Study in Taxonomic
      Theory." Journal of the History of Medicine
      24(1969): 294-309.

Describes John Vaughn Thompson's discovery in
1826 of larval barnacles and discusses the theoretical
problems this caused zoological taxonomy.

605. Winsor, Mary Pickard. Starfish, Jellyfish, and
      the Order of Life: Issues of Nineteenth-Century
      Science. New Haven: Yale University Press,
      1976.

Analyzes changes in classification from the
eighteenth century systems to those of the early
nineteenth century representing nature's order. Covers
from Cuvier to Darwin focusing on efforts to understand
Cuvier's radiata, the jellyfish, starfish, and polyps.

ZOOLOGY

606. Cole, Francis Joseph. The History of
      Protozoology. London: University of London
      Press, 1926.

Contains a very brief survey of the major names
and events in the history of protozoology.

607. Coleman, William. Georges Cuvier, Zoologist: A
      Study In the History of Evolution Theory.
      Cambridge: Harvard University Press, 1964.

Discusses the concepts and contributions of
Georges Cuvier beginning with comparative anatomy and
classification. Considers his approach to the problems
of species and explains his position on species fixity.

608.   Craigie, Edward Horne. A History of the
       Department of Zoology at the University of
       Toronto Up to 1962. Toronto: University of
       Toronto, Department of Zoology, 1966.

       Covers the development of the Department from its
inception in 1853 to 1962, identifying the individuals
and important dates connected with it. Specifies the
mid 1930s as the transition from morphology to
experiment.

609.   Davenport, Guy. The Intelligence of Louis
       Agassiz: A Specimen Book of Scientific Writings.
       Boston: Beacon, 1963.

       Contains an historical introduction and reprints
excerpts from five of Louis Agassiz's works, each with a
descriptive note accompanying it.

610.   Dexter, Ralph W. "Historical Aspects of F.W.
       Putnam's Systematic Studies on Fishes." Journal
       of the History of Biology 3(1970): 131-135.

       Describes the promising early icthyological
career of Frederic Ward Putnam, who turned from
icthyology to archeology.

611.   Gillespie, Neal C. "Preparing for Darwin:
       Conchology and Natural Theology in Anglo-
       American Natural History." Studies in History
       of Biology 7(1984): 93-145.

       Describes the development alongside natural
theology of scientific practices and explanation in the
field of conchology in the first half of the nineteenth
century. Discusses issues raised by taxonomy,
geographical distribution, creationism, evolution,
positivism, professionalism, and natural theology.

612.   Groeben, Christiane, ed. Charles Darwin, 1809-
       1882, Anton Dohrn, 1840-1909: Correspondence.
       Naples: Gaetano Macchiaroli Editore, 1982.

       Publishes correspondence between Charles Darwin
and the Prussian biologist Anton Dohrn, founder of the

Naples Zoological Station. Dohrn's research and the Station itself were stimulated by the Darwinian theory. Dohrn initiated the correspondence with Darwin in 1867; it continued until 1882 and Darwin's death.

613. Groeben, Christiane, and Irmgard Muller. The Naples Zoological Station at the Time of Anton Dohrn: Exhibition and Catalogue. Naples: Stazione Zoologica di Napoli, 1975.

Records the history of the Naples Zoological Station in a brief history by Irmgard Muller and through the entries written for an extensive and fascinating exhibit on the history and contributions of the Station. Includes photographs and a comprehensive bibliography of articles about the Station starting with Anton Dohrn's earliest in 1871.

614. Hall, Thomas Steele, ed. A Sourcebook in Animal Biology. New York: McGraw-Hill, 1951.

Reprints articles, chapters, and excerpts, in translation where necessary, of classic works in animal biology from the Renaissance to the end of the nineteenth century. Covers the field broadly, including comparative and systematic zoology, physiology, behavior, embryology, cellular biology, pathology, evolution, and zoogeography. Includes brief comments on the reprinted material.

615. Harvey, E. Newton. A History of Luminescence from the Earliest Times until 1900. Philadelphia: American Philosophical Society, 1957.

Contains a 130 page section on the history of observations and experiments on bioluminescence, the luminescence of living things, including a chapter on animal luminescence that treats fireflies and glowworms from the seventeenth century beliefs through the nineteenth century physiological and histological studies.

616. Locy, William A. The Main Currents of Zoology. New York: Henry Holt and Co., 1918.

Covers all of the history of zoology at an introductory popularized level; remains the only attempt at comprehensive treatment.

617.   Lurie, Edward. <u>Louis Agassiz. A Life in Science</u>. Chicago: University of Chicago Press, 1960.

Discusses Agassiz's education and international career in the rich context of the nineteenth century world of natural history.

618.   Petit, G., and Jean Théodoridès. <u>Histoire de la Zoologie du Origines a Linné</u>. Paris: Hermann, 1962.

Covers the history of the identification, description, and classification of animals.

619.   Théodoridès, Jean. <u>Un Zoologist de L'Epoque Romantique: Jean-Victor Audouin (1797-1841)</u>. Paris: Bibliothèque Nationale, 1978.

Describes the life and zoological work of Jean Victor Audouin.

AUTHOR INDEX

144 Author Index